THE DIVINE WORD
AND THE GRAND DESIGN

THE DIVINE WORD
AND THE GRAND DESIGN

Interpreting the Qur'an
in the Light of Modern Science

Basil Altaie

First published in the UK by Beacon Books and Media Ltd
Innospace, Chester Street, Manchester M1 5GD, UK.

Copyright © Basil Altaie 2019

The right of Basil Altaie to be identified as the author of this work has been asserted in accordance with the Copyright, Designs and Patents Act 1988. All rights reserved. This book may not be reproduced, scanned, transmitted or distributed in any printed or electronic form or by any means without the prior written permission from the copyright owner, except in the case of brief quotations embedded in critical reviews and other non-commercial uses permitted by copyright law.

First paperback edition published in 2019

www.beaconbooks.net

ISBN: 978-1-912356-19-5 Paperback
ISBN: 978-1-912356-20-1 Hardback

Cataloging-in-Publication record for this book is available from the British Library

Cover Image: Klemen Vrankar, unsplash.com

Credits:
Every effort has been made to trace copyright holders and to obtain their permission for the use of copyright material. The publisher apologises for any errors or omissions in the above list and would be grateful if notified of any corrections that should be incorporated in future reprints or editions of this book.
Fig 1.Pearson Education; Fig 3 National Aeronautics and Space Administration; Fig 9 Montana State University; Fig 10 National Aeronautics and Space Administration; Fig 13 Roland Laffitte Héritage Arabe, Paris 2006; Fig 14 Russell Kightley; Fig 20 N F Gier, University Press of America

Contents

Foreword   vii

Chapter One
Qur'an: The Unique Book   1

Chapter Two
The Qur'an and Science   23

Chapter Three
The Solar System   43

Chapter Four
The Stars   79

Chapter Five
Time   87

Chapter Six
The Grand Design   91

Chapter Seven
Evolution   121

Chapter Eight
The Universe   147

Chapter Nine
Problematic Verses   175

Epilogue   189

Index   193

To Robin

Foreword

In this book, I intend to present my findings on what I regard to be the scientific signs in some verses of the Qur'an. I follow a new approach in discussing the scientific signs contained or alluded to in some verses of the Qur'an by subjecting these verses to the scrutiny of linguistic and scientific analysis. My main concern is with verses that contain astronomical or cosmological content, in which I deal with the subtle meaning of the words and the overall construction of the verse. This approach involves attesting to the formal presentation of the Qur'an with conjecture and verification, and requires one to be well-acquainted with the style and usage of metaphors in the Arabic language as well as the related scientific facts. Traditional commentators of the Qur'an were certainly well-acquainted with Arabic, including its meanings, usage and metaphorical style, despite lacking accurate knowledge of science. They were largely influenced by the knowledge and culture of their time and thus misinterpreted some verses, especially those which are dubious (*mutashābeh*).

Instead of following an apologetic approach in explaining the verses of the Qur'an and 'expose' its scientific signs, I have relied on two sources: the established scientific facts and the original Arabic meanings of the words as found in authentic lexicons. This makes the approach of verification much more reliable than presenting unfounded opinions or relying on weak narrations. For this reason, this book rarely points to narrations of the Prophetic hadith.

The first chapter of the book discusses the uniqueness of the Qur'an; it is no ordinary book, and if it was, it would have been forgotten long ago. In chapter two I discuss the relationship between the Qur'an and science, asserting the fact that although the Qur'an is no book of science, it contains remarkably precise wording expressing facts about man and nature in the most accurate way.

The subsequent chapters of the book consider some basic phenomena and prominent objects of the world such as the Sun, the Earth and the heavens by reflecting on the presentation of these in the Qur'an in order to testify to the scientific value of the information. In such an assessment there may be differences of opinion, which admittedly may all appear plausible at first glance. However, once we put these views under scrutiny, taking into consideration the wording and the available meanings in Arabic, we can distinguish the correct one.

A chapter is devoted to discussing the creation, development and fate of the universe. This topic again contains some dubious verses that can cause contention since the meaning of the words 'heaven' and 'heavens' have several facets. Analysing such a quantity of verses requires serious attention and hard work, since these two words have been repeated in the Qur'an 310 times and appear in different contexts. Moreover, there may appear to be a contradiction between the views expressed in the Qur'an about the fate of the universe—telling us that the universe is going to collapse—and the most recent scientific discoveries which suggest that the universe will go on expanding forever. A scientific explanation is given using the most modern observations in cosmology, presenting support for the idea of a possible collapse of the universe.

In the last chapter of this book I discuss some of the 'problematic verses' in the Qur'an and attempt to show that the problems raised by such verses can be solved, in most cases, by considering the wording of the Qur'an and reflecting on the available meanings in Arabic.

Finally, I would like to thank Dr. Ahmed G. Hamam for reading the book and offering some suggestions which have contributed greatly to its clarity.

Dr. Basil Altaie
Professor of Physics and Cosmology
Yarmouk University, Jordan

Leeds, September 2018

Chapter One

Qur'an: The Unique Book

Muslims believe that the Qur'an is the word of Allah, the Creator and Sustainer of the world. Readers of the Qur'an in Arabic, Muslim or non-Muslim, may quickly recognise the special character of this book. It is a character that is built-in through the structure, choice of words, construction of phrases, and presentation of verses.

The Qur'an contains the history of certain social events that took place during the time of the Messenger ﷺ. It tells stories of messengers and prophets that came before Prophet Muhammad ﷺ, such as Adam, Noah, Abraham, Isaac, Jacob, Joseph, Moses, Jesus and many others.

The Qur'an stipulates the laws of a Muslim's life, rituals and morals; a law that is known as the *shari'ah*. It is unlike any book or any other scripture in structure and content in that it cannot be subjected to hermeneutical analysis because of the unconventional structure of its construction. This is what makes the Qur'an a unique text. It also contains many statements that point to natural objects and phenomena like the Sun, the Moon, the planets, the day and the night, the rain, meteors, the growth of plants, rivers, seas, mountains, trees, earthquakes, and the development of the foetus in the womb. Some of these are described clearly, but others are introduced through metaphoric expressions.

The Divine Word and the Grand Design

The Qur'an contains no formal composition, except in a few places, when sharing specific stories like the story of the sons of Israel. Otherwise, the narrative is in most cases discontinuous. This discontinuity is meant to entangle the Qur'an's phrases and verses in a rotary planar style, pointing always to the essence of the Islamic creed: the *Tawhid*.

I claim the Qur'an contains proof of its own authenticity, and that these proofs are seen throughout its construction. First, there are several places which show that if Muhammad ﷺ or any human being was to write the Qur'an, he would have certainly written some parts very differently. Second, in many cases, the choice of words in the Qur'an is made to reflect more than one meaning, allowing for different levels of understanding. It is remarkable that none of these meanings are in mutual contradiction. These interpretations do not lead to misguidance since there is always one word or more that act as a signpost to direct the reader to a higher level of understanding. Third, the construction of the verses usually contains tailing; a statement that directs the reader to follow up on the issue with which the verse is concerned. For example, the verse [13:3], which tells us about how the land is spread, how rivers are laid and how crops are twined, is followed by a sentence that reminds people of the succession of day and night. This seems to imply the rotation of the globe when seen in the light of other related verses. Other verses encourage the reader to contemplate the content of the verse. For example, in [16:11] and [45:13] we find the tailing: *There is a sign in this for a people who reflect.* Other tailing of verses addresses the mind of the reader and incites him or her to understand what the verse is pointing to, for example: [2:164], [13:4] and [16:67].

In some verses, knowledgeable people are commanded to see the proof contained therein, specifically in places where delicate information is revealed. For example, [32:30] and the verse [29:43] reads: *We cite these examples for the people, and none appreciate them except the knowledgeable.* The context of this verse is the mention of how subtle and weak the spider's web is, being the frailest of all dwellings. Then the Qur'an states: *If only they knew it.*

This tailing of the verse is stunning. Why should trivial information (in this case, the weakness of the spider's web) be pointed to as if it is something mysterious? Everybody knows the spider's web is extremely weak. Certainly, the verse is pointing to something beyond the spider's web. The implication is understood if we read the verse in Arabic and see that it actually mentions the 'home' (*bayt*) of the spider, not merely the web. Home is not just a house. Home involves beings and relationships among those beings. Animal behaviour specialists know much about the weakness of family relationships between spiders.

The very first words revealed in the Qur'an encourage people to read in the enlightenment of the revelation, and seek knowledge through the creation and through contemplation.

Read! In the Name of your Lord, Who has created (all that exists).
[96:1]

{اقْرَأْ بِاسْمِ رَبِّكَ الَّذِي خَلَقَ} [العلق:1]

Also, we can see that the second *surah* revealed to Muhammad ﷺ mentions the Pen as a means of scribing and documenting knowledge, a motivation for people to learn how to write.

The Qur'an states that the purpose behind pointing to natural objects and events is to attract attention to the greatness and glory of the Creator who made all of it possible. In fact, the Qur'an motivates believers to journey through the Earth and search for how creation began.

Say: Travel in the land and see how (Allah) originated creation, and then Allah will bring forth (resurrect) the creation of the Hereafter (i.e. resurrection after death). Verily, Allah is Able to do all things. [29:20]

{قُلْ سِيرُوا فِي الْأَرْضِ فَانْظُرُوا كَيْفَ بَدَأَ الْخَلْقَ ثُمَّ اللَّهُ يُنْشِئُ النَّشْأَةَ الْآخِرَةَ إِنَّ اللَّهَ عَلَى كُلِّ شَيْءٍ قَدِيرٌ} [العنكبوت:20]

The Divine Word and the Grand Design

When scanning through the Qur'an, we see there are phrases that address simple observations in nature, questioning the power behind creation and the design of the world. For example, the Qur'an asks:

> *Do they not look at the camels, how they are created? And at the heaven, how it is raised? And at the mountains, how they are rooted and fixed firm? [88:17–19]*

{أَفَلَا يَنْظُرُونَ إِلَى الْإِبِلِ كَيْفَ خُلِقَتْ (17) وَإِلَى السَّمَاءِ كَيْفَ رُفِعَتْ (18) وَإِلَى الْجِبَالِ كَيْفَ نُصِبَتْ} [الغاشية:17-19]

Such questions address the layman and describe facts in a simplistic way. This might have been enough to challenge those who were ignorant of the Creator. The aim is clear: the question here is to attract the attention of laymen and to have them acknowledge the need for a Creator and Supreme Power.

In other verses, we find more serious questions that address different kinds of people, such as those who are knowledgeable, and people of science and wisdom. Such questions address much more complicated topics, posing challenges to encourage deep thought. For example:

> *Say: Have ye thought, if Allah made night everlasting for you till the Day of Resurrection, who is a God beside Allah who could bring you light? Will ye not then hear? [28:71]*

{قُلْ أَرَأَيْتُمْ إِنْ جَعَلَ اللَّهُ عَلَيْكُمُ اللَّيْلَ سَرْمَدًا إِلَى يَوْمِ الْقِيَامَةِ مَنْ إِلَهٌ غَيْرُ اللَّهِ يَأْتِيكُمْ بِضِيَاءٍ أَفَلَا تَسْمَعُونَ} [القصص:28:71]

Also, the next verse says:

> *Say: Have ye thought, if Allah made day everlasting for you till the Day of Resurrection, who is a God beside Allah who could bring you night wherein ye rest? Will ye not then see? [28:72]*

{قُلْ أَرَأَيْتُمْ إِنْ جَعَلَ اللَّهُ عَلَيْكُمُ النَّهَارَ سَرْمَدًا إِلَى يَوْمِ الْقِيَامَةِ مَنْ إِلَهٌ غَيْرُ اللَّهِ يَأْتِيكُمْ بِلَيْلٍ تَسْكُنُونَ فِيهِ أَفَلَا تُبْصِرُونَ} [القصص:72]

Qur'an: The Unique Book

The question here is: how is it possible for Allah to make the night everlasting? And how is it possible to make the day everlasting? Would it have to be a miracle? Wouldn't that violate the laws of nature and the divine *sunnah*? Furthermore, and more importantly, why did the Qur'an state both verses (the night in one verse and the day in the next)? Wouldn't it be sufficient for the challenge to state just one of them?

In another verse, the Qur'an asks the question:

> *Do you see how thy Lord extends the shade? And if He willed, He would have made it stationary, then We have made the Sun a guide for it. [25:45]*

{أَلَمْ تَرَ إِلَى رَبِّكَ كَيْفَ مَدَّ الظِّلَّ وَلَوْ شَاءَ لَجَعَلَهُ سَاكِنًا ثُمَّ جَعَلْنَا الشَّمْسَ عَلَيْهِ دَلِيلًا} [الفرقان:45]

The challenge in this verse is: how can Allah make the shade stationary without violating the laws of nature? This will be discussed later in the book.

The Qur'an says knowledgeable people are those who know this book is telling the truth:

> *And those who have been given knowledge see that what is revealed to you from your Lord is the truth, and guides to the Path of the Exalted in Might, Owner of all praise. [34:6]*

{وَيَرَى الَّذِينَ أُوتُوا الْعِلْمَ الَّذِي أُنزِلَ إِلَيْكَ مِن رَّبِّكَ هُوَ الْحَقَّ وَيَهْدِي إِلَى صِرَاطِ الْعَزِيزِ الْحَمِيدِ} [سبأ:6]

There are many places where the Qur'an presents facts about the natural world using thought-provoking words that makes one feel the revelation is from a divine source. These places explain the meaning of the phrases and state that verses of the Qur'an are proofs (*ayāt*) for its authenticity as the word of Allah.

The Prophet Muhammad ﷺ performed no famous miracles, as was the case with Moses and Jesus, but he has been supported by the greatest of all miracles: the perpetual miracle that is the Qur'an. Just like Jesus is given the title 'Word of Allah', signifying revelation, so is the Qur'an.

The Prophet Muhammad ﷺ had no influence in casting the words of the Qur'an—a fact that we see as we go through the verses with an open mind and read the content consciously and impartially, with due consideration for the Arabic construction involved. The Qur'an exposes this by surveying topics related to astronomy and cosmology. Throughout the chapters of this book, I will expose places which prove that neither Prophet Muhammad ﷺ (nor any other human) could have written such a book—whether they lived 1400 years ago or in any other time period. Our current scientific knowledge helps us discover clear indications of the divine authenticity of the Qur'an.

Of all the known holy books of the monotheistic religions, the Qur'an is unique in its way of preservation. Whenever a verse or several verses were revealed to Prophet Muhammad ﷺ, he called designated scribes from his followers and told them to write down what was revealed to him. Then, he ﷺ would ask them to recite what they had written, and he himself would recite these verses to followers during the five daily prayers. This enabled a large number of his followers to memorise the entire Qur'an.

After the death of the Prophet ﷺ the number of those who had memorised the Qur'an began to decrease, so prominent followers, under the auspices of Caliph Othman and the supervision of Imam Ali Ibn Abi Talib, arranged to prepare one unified text of the Qur'an, which was copied and distributed to the Islamic states at the time.

The Qur'an and the Mind

Life as we live it has more than one dimension. We are deluded by the needs of our body; eating, drinking, procreating, sleeping and the rest of our natural desires and needs. When performing these activities, we behave like high-ranking animals. This is the natural dimension. But as we reflect upon our comprehension and environment, we soon find that there are more dimensions to our life. We become curious about events; we have a desire to ask why this happens and not that.

Where did it come from? How can this work? Why are we born like this? Is it not a fact that we have many things in common with animals, such as morphology, anatomy and functions? Even our psychology is somewhat similar to that of high-ranking animals. Such curiosity brings us to the dimension of rationality, leading us to logic and scientific enquiry.

But once we go through our scientific enquiry, we find there are things that cannot be explained rationally. Everywhere we look, we face a dead end. There is always that agony of questioning: but why? The whole universe stands before us as an open-ended jigsaw puzzle. There will never be a final answer to our questions. Science has no end, no final answer, and despite our confidence and arrogance, there will be no final theory of everything—never! But why?

This drives us to suspect there must be another dimension of life that rational comprehension may not be able to uncover: a spiritual, irrational dimension. But how can we comprehend irrationality through our rational comprehension? How can we get to the state where we can appreciate values beyond the observed properties of things? How can we go beyond our natural world when we have been constructed in a way that obeys the laws of nature, and when the laws themselves are only concerned with the materialistic world? In this age, this is the greatest challenge our consciousness faces.

During the last four centuries, scientific development has enabled us to probe into new understandings and construct new visualizations of the world that can take us beyond the boundaries of our direct sensation. This requires us to realise a new dimension, one that is induced by our rational contemplations.

This dimension may answer some fundamental questions that animals may not ask: Why are we here? What is the purpose of our existence,

The Divine Word and the Grand Design

what is this world, why should it be so, what is our destiny? Is our existence a mere chance? History of science tells us that we have assumed events happen by chance, only to later discover that such events have causes. Chance is the materialistic analogue of the God-of-the-gaps which is sometimes employed by naïve religious apologetics. It is difficult for our minds to comprehend that the universe is governed by a transcendental being. Many great minds could not grasp the idea of a 'super-being', although they addressed its existence in several ways. Pierre Laplace acknowledged that if such a being does exist, then this agency could fully control the past, the present and the future. He says "An intelligence knowing all the forces acting in nature at a given instant, as well as the momentary positions of all things in the universe, would be able to comprehend in one single formula the motions of the largest bodies as well as the lightest atoms in the world, provided that its intellect were sufficiently powerful to subject all data to analysis; to it nothing would be uncertain, the future as well as the past would be present to its eyes. The perfection that the human mind has been able to give to astronomy affords but a feeble outline of such intelligence."[1]

Indeed, rationally it could be difficult to comprehend that an agency beyond the universe could sustain such a complex system. It would be inconceivable that such intelligence could control every atom and every electron in the universe. Albert Einstein believed that the whole order in the universe constitutes of an agency of control and development. For him, the difficulty of accepting the God of major religions, particularly monotheistic religions, stems from his inability to accept the notion of a personal God that rewards and punishes.[2] Indeed, the notion of a personal God as imagined by classical religions is difficult to grasp rationally—this is the most difficult challenge for human beings.

---

1 Laplace, Pierre. *A Philosophical Essay on Probabilities*. Translated by F. W. Truscott and F. L. Emory. New York: Dover, 1951.

2 Max Jammer, Einstein and Religion, Princeton University Press, 1999.

The Falsifiability of Science

According to Karl Popper, scientific knowledge should be falsifiable. For this reason, he believed religious knowledge can be conceptualized as 'belief' but not 'science'. The biggest problem with religious and spiritual knowledge is the challenge of verification as in most cases, science is verifiable. This is one of the major differences between faith and science.

However, following on from this idea, if we must adopt the principle that scientific knowledge is always falsifiable then the truthfulness of such knowledge is always at stake. As such, the scientific knowledge will be inherently doubtful and characterized as true in essence, though it marks a lack of truth and weakness of credibility. If scientific knowledge is characterized with the requirement of falsifiability, then we may proceed to consider faith as being more credible once it is subjected to verifiability. Verifiable beliefs are then more credible than falsifiable science. For this reason, I believe we need other criteria by which we can bridge science and belief. In one sense, we are able to identify scientific knowledge with belief. Towards the end of the 19th century, Newtonian mechanics and the theory of universal gravitation was taken as belief despite being supported by many observations which confirmed the accurate calculations of positions and periods of astronomical objects. Such observational verifiability confirmed the truth of those laws; so why not adopt verifiability as a measure for the truth of our beliefs in order to enable us to compare it with falsifiable scientific knowledge? This seems to be possible even if we cannot provide rigorous theoretical justification for religious belief. There should always be enough logical justification to take a specific belief as verifiable. In the Qur'an, the words and phrases are considered proofs (ayāt) and it is this power of verifiability that gives strength to the argument.

The Divine Word and the Grand Design

Dubious Verses

It should be noted that the Qur'an has stated that some of its verses are 'unclear' or 'confusing'. These I call dubious (*mutashābeh*) verses. In Surah Al-Imran we read:

> It is He Who has sent down to you (Muhammad) the Book. In it are Verses that are entirely clear, they are the foundations of the Book; and others not entirely clear. Those whom in their hearts there is a deviation would follow that which is not entirely clear thereof, seeking confusion among people, and seeking for its hidden meanings, but none knows its hidden meanings save Allah. And those who are firmly grounded in knowledge say: "We believe in it; the whole of it is from our Lord." And none receive admonition (grasp the meanings of the confusing verses) except those of understanding. [3:7]

{هُوَ الَّذِي أَنْزَلَ عَلَيْكَ الْكِتَابَ مِنْهُ آيَاتٌ مُحْكَمَاتٌ هُنَّ أُمُّ الْكِتَابِ وَأُخَرُ مُتَشَابِهَاتٌ فَأَمَّا الَّذِينَ فِي قُلُوبِهِمْ زَيْغٌ فَيَتَّبِعُونَ مَا تَشَابَهَ مِنْهُ ابْتِغَاءَ الْفِتْنَةِ وَابْتِغَاءَ تَأْوِيلِهِ وَمَا يَعْلَمُ تَأْوِيلَهُ إِلَّا اللَّهُ وَالرَّاسِخُونَ فِي الْعِلْمِ يَقُولُونَ آمَنَّا بِهِ كُلٌّ مِنْ عِنْدِ رَبِّنَا وَمَا يَذَّكَّرُ إِلَّا أُولُو الْأَلْبَابِ} [آل عمران:7]

Firstly, we should recognise that most of the dubious verses are concerned with information related to natural objects or natural phenomena. During the period in which the Qur'an was revealed, people were not acquainted with accurate knowledge concerning the creation of the world. Consequently, if they were told the correct and accurate facts, they may have become more confused and would proclaim that the Qur'an is nonsensical. Such an attitude on behalf of the public would have hindered the propagation of the message of Islam. Incidentally, this may also apply to people during various stages of history since our knowledge about the natural world, including ourselves, is never complete. Therefore, understanding or interpreting the dubious parts of the

Qur'an cannot truly be achieved. It is for this reason that the Qur'an states: *but none knows its hidden meanings save Allah.*

Several books have been written explaining the dubious verses but I could not find a clear and convincing argument for the existence of such verses in the first place. Some authors believe these verses exist to test the believers and differentiate them from the non-believers or apostates. This opinion is based on the Qur'anic verse: *Those whom in their hearts there is a deviation would follow that which is not entirely clear thereof.* But why should Allah set a trap for those who are deviating from the right path with verses that are inherently dubious? For this reason, I find this interpretation unconvincing. This verse tells us that those who deviate from the straight path would use the unclear meanings as a chance to spread confusion amongst people, whereas those who believe and have knowledge would not look for further interpretation. This brings forth the question: why do these verses exist?

I argue that unclear verses in the Qur'an exist because the Qur'an is the word of Allah and is derived from His absolute, precise and comprehensive knowledge. Meanings become unclear once this absolute knowledge is confined in words and phrases in the structure of language (Arabic). We can see this in the example of the words 'heaven' and 'heavens'. In a research article[3] my colleague and I analysed these terms, which were found in many verses in the Qur'an. The meanings of these terms differ depending on the context. We sought meanings for these words using astronomical knowledge and found that there are several meanings for the word 'heaven' (single), such as the firmament, the sky, the celestial sphere above us and the entire universe. But the word 'heavens' (plural) was harder to interpret. I will discuss this in more detail in the final chapter of the book.

However, it should be noted that the ability to interpret the dubious verses cannot be achieved without an extended knowledge of both

---

[3] M.B. Altaie and M.K. Alzubi, "The Concept of Heaven and Heavens in the Qur'an and Modern Astronomy", Jordanian Journal of Islamic Studies, vol. 4, No.3, p. 223-249. 2008.

science and the Arabic construction of the Qur'an. The development of our understanding of scientific knowledge may help one to interpret the Qur'anic verses with unclear meanings. Such verses can be comprehended over time, although some deeper understanding will remain concealed. This belief is supported by the narration of the Prophet where he is reported to have said, "We, the Prophets, are sent to talk to people with what they can comprehend." For a devout Muslim, the fundamental element of belief is to acknowledge that the Prophet Muhammad ﷺ is a messenger of Allah. This is why the testimony of faith contains the belief that Allah is one and that Muhammad is His messenger.

Faith

The Islamic faith is centred on Tawhid. The testimony of Tawhid is to acknowledge that 'There is no God but Allah'. This testimony is not mere words, but rather constitutes a whole belief and an everyday practice that is achieved when one becomes free of any external pressure or obligation to any creature other than Allah. With this understanding, Tawhid can be considered the total liberation of a human being from the authority of anyone except Allah. It means that the world is one and the Creator is one, and helps us to establish harmony with the rest of the world as we are part of the same creation.

Through this comes the dimension of 'faith' or 'belief', which contains several facets: the first is to accept certain ideas passionately without rational analysis, and the second is to interact enthusiastically with situations or thoughts and react accordingly. This aspect of faith allows us to experience the beauty of all things in life. It is the original source of love. It makes us feel that we are connected to the universe and share a common belonging. This is not limited to religious people, as the astronomer Carl Sagan has expressed a similar sentiment. These dimensions contribute to our life—from sorrow and joy, optimism and pessimism, to hopes and frustrations—and our lives flourish by our ability to experience feelings in a way that is distinct from any other creation.

Faith and the Heart

The Qur'an emphasises that learning can be acquired through the heart and not only through the intellect. The 'heart' is a metaphor to describe a consciousness that goes beyond literal meanings and sees the invisible, and seeks the hidden meaning behind material things. This is known as semiotics. As described in the Qur'an, semiotics is of two types: rational and spiritual. The rational is attained by analysing signs and construing their implications, with the help of our previous knowledge. It is a kind of hermeneutical analysis that leads to obtaining new information by connecting the sign with other related facts, terms or images, for example, a fountain may be related to flourishment and fertility. Spiritual semiotics involves the contemplation of signs through our conscious interaction with the symbol, involving our emotions expressed toward it. Spiritual semiotics also has a further level: 'revelation'. During this the content is understood through the help of one of three enlightenments (*nūr*), listed in order: the divine enlightenment (*nūr ilahi*), the Qur'an's enlightenment (*nūr Qur'ani*) and the Prophet's enlightenment (*nūr Muhammadi*). Any one person is able to obtain one or more of these enlightenments should they follow the correct path to Allah, by understanding the Qur'an and following the instructions of the Prophet ﷺ.

Divine enlightenment is a gift from Allah. It is for Him to decide who deserves it and for what purpose. In his book, *The Deliverance from Error (Al-Munqith min al-Dalal)*, Abu Hamid al-Ghazali describes how Qur'anic enlightenment is attained by a careful reading of the Qur'an and contemplating the meaning of its phrases. In *The Jewels of the Qur'an (Jawahir al-Qur'an)*, he goes on to write about how Qur'anic enlightenment can be acquired by mastering the language and becoming involved with the spirit of the Qur'an. We receive Prophetic enlightenment through the Prophet Muhammad ﷺ by following his instructions and deeply absorbing his morals and ethics. The sahaba (companions of the Prophet) attained this superior enlightenment by being near to him, living and working with him.

In Arabic the word nūr means 'light', but it is not quite the light we observe physically, as it is associated with the heart and not the eye. The eye senses visible light and the heart senses nūr. Nūr is not a physical object but a metaphor to describe enlightenment that is given to the mental and psychological construct that recognises the unseen. This is an important factor in understanding the Qur'an, as shown in the verse from Surah al-Aʿraf:

> *And if you call them to guidance, they hear not and you will see them looking at you, yet they see not. [7:198]*

{وَإِن تَدْعُوهُمْ إِلَى الْهُدَىٰ لَا يَسْمَعُوا وَتَرَاهُمْ يَنظُرُونَ إِلَيْكَ وَهُمْ لَا يُبْصِرُونَ}
[الأعراف:198]

This verse indicates that to 'look' through the eyes is not the same as to 'see'. Therefore, the non-believers who were looking at the Prophet Muhammad ﷺ could not truly see, as they did not acquire the heart that can recognise the enlightenment he brought. Incidentally, this includes every calling that contains a spiritual dimension.

This also applies to natural events and objects. Some people can see the symbolic character of events and objects through which they can recognise purpose and values beyond material constructs. Others may not see anything beyond material content. For example, if one looks at an Impressionist painting, they may see it as a collection of colours, or nothing more than geometrical shapes and regular or irregular patterns. They may identify that there is a sun, a tree and a river. Although they see the colours of the red-painted river, the black Sun and the yellow tree, they are unable to recognise any symbolic meaning (apart from assuming the painter suffers from colour blindness!). Another person may see that the painter is describing a world full of evil and injustice; the river painted in red as the colour of blood, and the Sun being black reflecting that it is dull, and the yellow tree as a sign of dryness and impotency.

Allah the Sustainer

A key tenet of the Islamic creed is the belief that Allah is the Sustainer of the world, which is also emphasised in the Qur'an. He is the Omniscient, the Omnipotent and the One who has knowledge of everything and is capable of doing anything He wills. This belief may seem easy to accept as a part of faith, but is difficult in practice. We see that natural phenomena take place regularly and reliably, such that it seems to happen deterministically whenever the related conditions are made available. At first glance, one cannot see the role of a creator or sustainer as the world appears self-sustainable via the 'laws of nature.' Thus, it may seem that belief in God exists as a superstition or to fulfill a psychological need.

In several verses the Qur'an states that Allah has created the world with truth:

> *He it is Who made the Sun a shining brightness, and the Moon a light, and ordained for it stages that you might know the computation of years and the reckoning. Allah created not this but with truth. He makes the signs manifest for a people who know. [10:5]*

{هُوَ الَّذِي جَعَلَ الشَّمْسَ ضِيَاءً وَالْقَمَرَ نُورًا وَقَدَّرَهُ مَنَازِلَ لِتَعْلَمُوا عَدَدَ السِّنِينَ وَالْحِسَابَ مَا خَلَقَ اللَّهُ ذَٰلِكَ إِلَّا بِالْحَقِّ يُفَصِّلُ الْآيَاتِ لِقَوْمٍ يَعْلَمُونَ} [يونس:5]

> *Do you not see that Allah has created the heavens and the Earth with truth? If He wills, He can remove you and bring (in your place) a new creation! [14:19]*

{أَلَمْ تَرَ أَنَّ اللَّهَ خَلَقَ السَّمَاوَاتِ وَالْأَرْضَ بِالْحَقِّ إِنْ يَشَأْ يُذْهِبْكُمْ وَيَأْتِ بِخَلْقٍ جَدِيدٍ} [إبراهيم:19]

> *He has created the heavens and the Earth with truth. High be He*
> *Exalted above all they associate as partners with Him. [16:3]*

{خَلَقَ السَّمَاوَاتِ وَالْأَرْضَ بِالْحَقِّ تَعَالَىٰ عَمَّا يُشْرِكُونَ} [النحل:3]

For 'truth' to be measured requires known rules that are verified. This implies the presence of laws with which Allah has created and formed the world. Every creation is subject to specific calculations and each element is precisely allocated. Indeed, without this we cannot recognise the meaning of the truth by which creation happened. In a later chapter I will discuss whether the existence of the laws of nature supersedes divine intervention.

Faith and Eternity

It is difficult to comprehend the fact that we become mere dust at the end of our lives. The belief that the soul exists beyond our direct consciousness and beyond the chemicals which compose our bodies helps one to recognise that although our bodies may turn to ash, our souls do not cease to exist. The resurrection of the body in the afterlife is a fundamental component of the Islamic faith, without which there is no meaning or value for our life in this world. Muslims believe that all people will be resurrected on the Day of Judgment and will receive an outcome of their state in the eternal hereafter. The details of this day are theologically speaking, controversial. However, modern science can contribute towards our religious understanding of some of the general features of resurrection. This will be discussed in the final chapter on the fate of the universe.

In order to rationally understand the existence of the soul, we ought to reflect on our consciousness, which allows us to recognise we have a kind of transcendent extension. Rene Descartes identified that body (matter) is characterized by spatial extension and motion, while the mind is characterized by thought. Before Descartes, several Muslim scholars considered the subject of consciousness and showed how the human being receives spiritual knowledge through the elevation of their consciousness into a transcendent level, which allows them to be integrated with the rest of the world. This is aligned with the mystic (*Sufi*) spiritual experience that I too have experienced over several stages of my own life. Some of these were not merely illusions, but real events which occurred beyond my normal everyday experience.

From a reductionist point of view, transcendence should be reducible into materialistic action or psychic level of excitation. Such an understanding hinders obtaining a successful model for human consciousness. This by no means suggests that we should withdraw from rational thinking and scientific methodology; rather, we ought to extend our vision of scientific verification by employing whatever experience we have of the hidden world of reality. For example, it is common knowledge that imaginary numbers play a very important role in physics. Quantum tunneling would not be possible without the use of imagination for the wave function. We also know that the comprehensive structure of the world as a whole is composed of two parts: the time-like world which is causal, and the space-like world which is non-causal, both of which are separated by the light-like world, i.e. the wall of light. Why don't we employ these ideas in order to explain consciousness? Is it because we want to explain the world using only that which links to the time-like world, or is it because we are actually ignorant of the physical value of the non-measurable quantities?

I believe that we can explain consciousness scientifically with a predictive theory if we broaden our conception of physical existence to cover the space-like world, including the pure imaginary world of tachyons and magnetic monopoles. Only then will we be able to understand how effects occur beyond the relativistic causality of Albert Einstein. This would lead to greater appreciation of the physical role of the scalar potential in James Clark Maxwell's electromagnetic theory and would result in an even better understanding of the entanglement of quantum states and the wholeness of the world.

Faith and Energy

Faith can bring you positive energy, enhance your performance and enable you to live a happy life. Put simply, faith gives people a feeling of security, relaxation and confidence. It leads to improved mental health as well as increased resilience in the face of life's difficulties. These are not merely subjective opinions, but supported by scientific discoveries in modern psychology and consciousness research.[4] The reason for not realising this fact comes from being ignorant of the value of faith in the midst of the materialistic rush of everyday life.

Prayer is a prime way of renewing faith and is the reason why Muslims are commanded to perform five prayers a day, allocated at specific astronomical times: dawn, noon, afternoon, sunset and night. However, these prayers are more than formal rituals; they create time and space to reconnect with the Creator and provide respite from the demands of daily life. In Islam, there is no mediator between Allah and man. Thus, it is important to establish a personal connection with Allah and direct your attention to the power sustaining the universe, and ask Him for help if you are in need. Through this spiritual practice, Muslims renew their faith multiple times a day.

From the physiology of the nervous system, we know that our brains and spinal cords are bathed in a fluid called Cerebrospinal Fluid (CSF). This clear, colourless fluid "acts as a cushion or buffer for the brain, providing basic mechanical and immunological protection to the brain inside the skull. The CSF also serves a vital function in the cerebral autoregulation of cerebral blood flow."[5] Interestingly, this fluid is known to turn over at a rate of 3–5 times a day; a similar number to the Islamic ritual prayers. Perhaps there is a connection between the two!

---

4 See for example: John Randolph Price, *The Planetary Commission: Planetary commission for Global Healing,* Quartus Books, 2017.

5 See Wikipedia: Cerebrospinal Fluid.

The Prophet Muhammad ﷺ and the Qur'an

The Qur'an was revealed to the Prophet Muhammad ﷺ by the angel Gabriel over a long period of time—approximately 23 years. These revelations were received by the Prophet with complete submission and were delivered to the people in its original form. The Prophet Muhammad ﷺ did not interfere in any way with the revealed messages and was warned not to deliver any content unless it was complete. He was chosen for the role of messenger for several reasons: his truthfulness, his justice, his mercy and morality. As such, the Qur'an praises the Prophet Muhammad ﷺ for his morals and describes him as a mercy for the people.

> *We have sent you for no other reason but to be a mercy for mankind. [21:107]*

{وَمَا أَرْسَلْنَاكَ إِلَّا رَحْمَةً لِلْعَالَمِينَ} [الأنبياء:107]

Authorship of the Qur'an

The Qur'an is characterized by being revealed in its literal content alongside its meanings, and this is what makes the Qur'an the authentic word of Allah. Being written by scribes immediately under the supervision of the Prophet himself makes it the most authoritative and reliable scripture of all time.

It is important to note that the Qur'an itself prevents the Prophet Muhammad ﷺ from interfering in its content. The Prophet Muhammad ﷺ is asked to abstain from delivering any message from the Qur'an unless authorized. Such phrases of the Qur'an stand as additional evidence for the divine source of revelation:

> *Move not your tongue concerning (the Qur'an, O Muhammad) to make haste therewith. [75:16]*

{لَا تُحَرِّكْ بِهِ لِسَانَكَ لِتَعْجَلَ بِهِ} [القيامة:16]

The Divine Word and the Grand Design

Also, the Prophet Muhammad ﷺ was threatened with punishment if he delivered any message that was not a genuine revelation of the Qur'an:

> *It is a revelation from the Lord of the worlds. And if he (Muhammad ﷺ) had forged a false saying concerning Us (Allah). We surely should have seized him by his right hand. And then certainly should have cut off his life artery (Aorta). And none of you could withhold Us from (punishing) him. [69:43–47]*

{تَنزِيلٌ مِّن رَّبِّ الْعَالَمِينَ (43) وَلَوْ تَقَوَّلَ عَلَيْنَا بَعْضَ الْأَقَاوِيلِ (44) لَأَخَذْنَا مِنْهُ بِالْيَمِينِ (45) ثُمَّ لَقَطَعْنَا مِنْهُ الْوَتِينَ (46) فَمَا مِنكُم مِّنْ أَحَدٍ عَنْهُ حَاجِزِينَ} [الحاقة 43–47]

Infallibility of the Prophet Muhammad ﷺ

In several other verses, the Qur'an reprimanded the Prophet Muhammad ﷺ on issues where he took seemingly wrong decisions. The revelation immediately corrects the situation by revealing the correct decision, act or ruling. For example:

> *O Prophet! Why do you ban (for yourself) that which Allah has made lawful to you. [66:1]*

{يَا أَيُّهَا النَّبِيُّ لِمَ تُحَرِّمُ مَا أَحَلَّ اللَّهُ لَكَ} [التحريم:1]

In another verse the Prophet Muhammad ﷺ is instructed against defending some people who tried to deceive him by accusing an innocent person of theft:

> *And argue not on behalf of those who deceive themselves. Verily, Allah does not like anyone who is a betrayer of his trust, and indulges in crime. [4:107]*

{وَلَا تُجَادِلْ عَنِ الَّذِينَ يَخْتَانُونَ أَنفُسَهُمْ إِنَّ اللَّهَ لَا يُحِبُّ مَن كَانَ خَوَّانًا أَثِيمًا} [النساء:107]

The Qur'an has drawn attention to an incident when the Prophet Muhammad ﷺ did not pay attention to a blind man that came to him. The above verses and others lead us to understand the honour of being infallible. The fact that the Qur'an is pointing to these situations indicates that his infallibility is surely not an intrinsic one, but a result of the Divine correcting him through immediate revelations and instructing him to redirect his actions. In comparison to mankind, the Prophet ﷺ is infallible because there is an agent instantaneously correcting him, whereas there is no similar force correcting ordinary human beings. This, I believe, is the correct meaning of the infallibility of the Prophet Muhammad ﷺ.

> *He frowned and turned away. Because the blind man came to him.*
> *[80:1–2]*

{عَبَسَ وَتَوَلَّى (1) أَنْ جَاءَهُ الْأَعْمَى} [عبس:1-2]

The Qur'an warned the Prophet ﷺ that the enemy may try to confuse him in order to deliver a false revelation, but that he could be punished as a result.

> *They almost diverted you from the revelations we have given you. They wanted you to fabricate something else, in order to consider you a friend. If it were not that we strengthened you, you almost leaned towards them just a little bit. In that case We would certainly have made you to taste a double (punishment) in this life and a double (punishment) after death, then you would not have found any helper against Us. [17:73-75]*

{وَإِنْ كَادُوا لَيَفْتِنُونَكَ عَنِ الَّذِي أَوْحَيْنَا إِلَيْكَ لِتَفْتَرِيَ عَلَيْنَا غَيْرَهُ وَإِذًا لَاتَّخَذُوكَ خَلِيلًا (73) وَلَوْلَا أَنْ ثَبَّتْنَاكَ لَقَدْ كِدْتَ تَرْكَنُ إِلَيْهِمْ شَيْئًا قَلِيلًا (74) إِذًا لَأَذَقْنَاكَ ضِعْفَ الْحَيَاةِ وَضِعْفَ الْمَمَاتِ ثُمَّ لَا تَجِدُ لَكَ عَلَيْنَا نَصِيرًا} [الإسراء: 73-75]

The Divine Word and the Grand Design

This is indeed a very severe warning if it was written by the Prophet Muhammad ﷺ himself. The above verses affirm that the Prophet Muhammad ﷺ had no intervention in what was revealed to him.

The Prophet Muhammad ﷺ also used to recite the Qur'an loudly during the daily prayers and consequently, a large number of his followers memorised the Qur'an. This, besides the fact that the Qur'an was scribed under the supervision of the Prophet Muhammad ﷺ, helped greatly in preserving the holy text.

In our time, the Qur'an remains to be properly understood in the true spirit of its revelation. We need to understand the Qur'an in the spirit by which the Prophet Muhammad ﷺ and his followers understood it, and separate historical events from factual concrete instructions. We need to distinguish between revelations related to special circumstances or certain events and the general teaching of the Qur'an. This is needed in order to put everything in its proper context, be it religious commands or social regulations. Without this, there is the danger that it could lead to widespread confusion, inspire hatred, and cause injustice and tragedies. The Prophet Muhammad ﷺ was sent as a mercy to mankind. Our world now, more than ever before, is in need of the mercy and tolerance exemplified by the Prophet ﷺ in order to spread peace and harmony.

Chapter Two

The Qur'an and Science

It is said that four out of the five pillars of Islam require some scientific knowledge to be properly fulfilled: to determine the time of *salat* (five daily prayers), to calculate the *zakat* (almsgiving), to sight the crescent of Ramadan for the commencement of fasting and Shawwal for the end of fasting, as well as observing the crescent of Dhul Hijjah for *Hajj* (pilgrimage). Some knowledge of geography is also necessary in order to ascertain the *qibla* (direction of Mecca) during the prayers. However, I believe that all five pillars of Islam need science for the rational realisation of Tawhid (Unity of Allah).

The Qur'an asserts the belief in one Creator by positing rational arguments and asking people to acknowledge His domination over the world. However, it should be understood from the beginning that rational arguments in the Qur'an go beyond pure mental realisation and cross over into the realm of intuitive deduction. The type of logic found in the Qur'an primarily addresses the heart. It surpasses the formal logic of our consciousness to become a kind of sensation that is encountered by feeling rather than pure mental judgment. In this way, faith is characterized by both mental and emotional activities; both the logical and illogical.

According to the Qur'an, true knowledge and correct science consolidates belief in God, the Creator of the world. There can be no conflict between correct science and the Qur'an. If such a conflict arises then either our knowledge is incorrect or our understanding of the Qur'an is.

Muslim Cosmological Doctrines

Muslim cosmological doctrines and their approach to natural sciences are as closely bound to the metaphysical, religious, and philosophical ideas governing Islamic civilisation as the modern sciences are to the religious and philosophical ideas of the 16th and 17th century in the West. This close relation is best observed in the case of Muslim students who, upon the most cursory contact with the modern sciences, often lose their spiritual footing and no longer feel in harmony with their tradition, whereas the same students may have studied traditional mathematics and the natural sciences without being in any way alienated from the Islamic tradition. The facts of the sciences remain the same, but the difference lies in the manner and perspective in which they are interpreted. Consequently, the knowledge provided in the general Islamic vision of the cosmos is not only key for a true understanding of the Muslim sciences (and a necessary basis for any study of the history of medieval science), but also provides the principles which guide Muslims in judging all other natural sciences, since this knowledge is intrinsically bound with the immutable and non-historical essence of Islamic revelation. Only when the contours of the Islamic conception of the cosmos are clearly delineated will Muslims be able to absorb and integrate the elements of foreign sciences into their own worldview, in conformity with the spirit of their tradition.

This would require the establishment of a theo-rational basis as can be found in Islamic *kalam*, particularly in the principles of *daqiq al-kalam*.[6] A short exposition of these principles and their role in forming the

---

6 Basil Altaie, *Daqīq al-Kalām: the Islamic Approach to Natural Philosophy*, KRM Press. 2019.

Islamic approach to natural philosophy is given in my book *God, Nature and the Cause*.[7] With this approach, it may be possible to formulate a viable worldview that reflects the true Islamic attitude towards modern science. In fact, this is straightforward in terms of physics, since most of the doctrines of daqiq al-kalam align with the principles of modern physics. The areas that may need more effort to re-cast in an Islamic view are humanitarian subjects such as biology, sociology and ethics.

Natural Sciences and Revelation

In Islam, natural sciences are closely linked to revelation. The Islamic creed asserting the unity of Allah reflects the understanding of the unity of nature. There is a clear connection between the revelation and its recipients. As Seyyed Hossein Nasr avers, "It is much like that of the form to matter in the Aristotelean theory of hylomorphism. Revelation, or the Idea in its manifested aspect, is the form while the mental and psychic structure of the people who receive it acts as the matter upon which this form is imposed. The civilisation which comes into being in this manner, from the wedding of the above 'form' and 'matter', is dependent upon the psychic and racial qualities of the people who are its bearers in two ways. The first is that the Revelation is already spoken in the language of the people for whom it is meant, as the Quran insists so often; and, the second, that the 'matter' of this civilisation plays a role in its crystallization and further growth."[3]

Thus, unlike abstract sciences as mathematics and geometry, natural sciences are closely related to the perspective of the 'observer'. This makes natural sciences dependent on the qualitative essence of the civilisation in which they are cultivated. Within a single civilisation, different perspectives may exist in the study of the same topics. This is demon-

---

7 Basil Altaie, *God, Nature and the Cause*, KRM, 2016.

8 Seyyed Hossein Nasr, *An Introduction to Islamic Cosmological Doctrines*, Thames and Hudson, revised edition (1978), p.1.

strated through the variety of interpretations regarding the origins of the concepts and their philosophical implications. For example, the Ancient Greeks maintained that the regularity of a phenomenon is taken as an immutable law of nature that can never be breached. This produces, in general, a deterministic character for the phenomena, subsequently establishing the principle of deterministic causality. In traditional Islamic thought, the regularity of the natural phenomena is understood as a 'custom' of nature by which phenomena may abide, but a breach of this custom is always possible. In this way, causality is opened up to another interpretation; it becomes mere conjunction and loses its deterministic character.

The ancient and medieval cosmological sciences were based upon and attempted to demonstrate the basic doctrine of the unicity of nature. In previous civilisations, many did not consider there to be a relation between this doctrine and the unity of the Divine, but in Islam, the unicity of nature is a consequence of the unicity of the Divine. This is a fundamental difference from which several essential aspects of the Islamic worldview stem. The world is the creation of one God—Allah—and if it had been created by several Gods it would have become corrupted and ceased to exist.

> *Had there been other deities in the heavens and the Earth besides Allah, both the heavens and the Earth would have been destroyed. Glory to Allah, the Lord of the Throne above what they attribute to Him! [21:22]*

{لَوْ كَانَ فِيهِمَا آلِهَةٌ إِلَّا اللَّهُ لَفَسَدَتَا فَسُبْحَانَ اللَّهِ رَبِّ الْعَرْشِ عَمَّا يَصِفُونَ}
[الأنبياء:22]

In another verse, the Qur'an explains that the creation would be corrupted if there was more than one God:

> *Allah has not taken to Himself a son, nor is there with Him any (other) god—in that case would each god have taken away what he*

created, and some of them would have overpowered others. Glorified be Allah above all that they allege. [23:91]

{مَا اتَّخَذَ اللَّهُ مِنْ وَلَدٍ وَمَا كَانَ مَعَهُ مِنْ إِلَهٍ إِذًا لَذَهَبَ كُلُّ إِلَهٍ بِمَا خَلَقَ وَلَعَلَا بَعْضُهُمْ عَلَى بَعْضٍ سُبْحَانَ اللَّهِ عَمَّا يَصِفُونَ} [المؤمنون:91]

The unicity of nature therefore provides evidence for the unity of the Creator. In this way, the form of Islamic revelation directly led to the integration of the ancient sciences into Islam, as well as for the sciences cultivated in the Muslim world itself. The doctrine of the unicity of nature, which is based upon that of unity and relies on the essence and spirit of the form of revelation in Islam, is therefore the ultimate aim of all the sciences of nature. The degree to which a science succeeds in expressing this unicity is the criteria by which the success and validity of that science is judged.

Indeed, the identification of the role of revelation in Islam and its subsequent influence on society features prominently in the works of Muslim scholars and intellectuals throughout the Islamic Golden Age.

The Role of Arabic in Islam

Islam was revealed in the Arabic language to a people who were of the stock of the Semitic nomads, and later spread among other racial and ethnic groups without losing its original character. The centrality and importance given to nature in the Holy Qur'an cannot be overemphasised. The Qur'an continually refers to the phenomena of nature as signs of God to be contemplated by the believers. "The pre-Islamic Arabs to whom the Quran was first addressed had a great love for Nature and like all the nomads who wander endlessly in the great expanses of virgin Nature had a deep intuition of the presence of the invisible in the visible. Islam, which has always preserved the form of the spirituality of Semitic nomads, emphasised this particular trait of the nomadic spirit and made of Nature in Islam a vast garden in which the handiwork of the invisible gardener is ever present."⁹

"Another point emphasized in the Quran is that human reason, which is a reflection of the intellect, when healthy and balanced, leads naturally to acknowledge the unicity of the Divine rather than to a denial of the Divine and can be misled only when the passions destroy its balance and obscure its vision. When not impeded by external obstacles and set free to contemplate, reason, therefore, does not lead to rationalism in the modern sense of the word, that is, a negation of all principles transcending human reason, but becomes itself an instrument of Unity and a way of reaching the intelligible world."[10]

Nasr accurately remarks that, "Islamic art, instead of being 'rationalistic' as it might at first appear, leads the observer through the abstract symbols of geometry to the principle of Unity which can be represented only abstractly."[11] It was also this conception of reason that made the study of the mathematical sciences so widespread in the Islamic world, enabling Muslims to devise their theory of atomism and propose the abstract concept of un-extended atom[12] as a part of their worldview.

Arabic is the sacred language of Islam. Nasr recognises that "this language has not only a precision which makes it an excellent instrument for scientific discourse but also an inner dimension which enables it to be the perfect vehicle for the expression of the most esoteric forms of knowledge. This flexibility made it easy for the early translators to translate Greek, Syriac, Sanskrit, and Persian texts into Arabic, to coin new words with relative ease, and expand the meaning of already existing terms to include new concepts. The character of the Arabic language it-

---

9 Nasr, p. 6.

10 Ibid, p. 7

11 Nasr, *Doctrines*, p. 7.

12 Harry Wolfson was puzzled by the concept of unextendedness of the Islamic atom proposed by the mutakallimūn. He wondered how the Arabs could device such an abstract concept. (See: H. Wolfson, *The Philosophy of the Kalam*, Harvard University Press 1976, p.472–473).

self therefore was influential in the study of all the sciences in the Islamic world, including those concerning Nature."[13]

In the sciences of nature, the Muslims had a rich vocabulary of Arabic words to express all the diverse concepts and ideas connected with the cosmological sciences. The word 'nature' corresponds to the Latin *natura*, the Greek *physis* and the Arabic word *tabi'a* from the root (*tab'*), but has a somewhat different meaning in the classical languages.

Interestingly, the Qur'an explores its own relationship with the Arabic language. This extends beyond the formal linguistic expression of the words to the structure of its thought. This is demonstrated in the following verse:

> *Had We formed this as a Qur'an in a foreign language they would have said: "Why are not its verses explained in detail? What! (a Book) not formed in non-Arabic with Arabic content?"*[14] *[41:44]*

﴿وَلَوْ جَعَلْنَاهُ قُرْآنًا أَعْجَمِيًّا لَقَالُوا لَوْلَا فُصِّلَتْ آيَاتُهُ ءَأَعْجَمِيٌّ وَعَرَبِيٌّ﴾ [فصلت:44]

This suggests that if the Qur'an had been formed in the style, metaphor and structure of a language other than Arabic, there would have been difficulties in understanding the content and following the signs contained therein. Consequently, people would request further details and explanations. This calls attention to the deep connection between the Qur'an and the Arabic language. Indeed, as we will later see when discussing some of problematic verses of the Qur'an, Arabic expressions have been chosen carefully to provide more than one meaning, and it is only under the scrutiny of precise synonyms that the true meanings become apparent.

---

13 Nasr, p. 8.

14 This is my translation of the above verse. Most available translations do not make a distinction between *ja'alnāhu*, meaning formed, which is specifically in this verse and the word *anzalnahu*, meaning revealed, which comes in other verses.

The Divine Word and the Grand Design

Science from an Islamic Perspective

Is it possible to adopt an 'Islamic' attitude towards understanding the discoveries of natural sciences? One might say that science is an independent rational quest of nature and has no connection to religion. This may be true; however, we always need to explain the outcome of our observations and experiments and devise theories formulating what we observe in terms of a law, expressing the relationship between the variables related to the observed phenomena. In this aspect of scientific endeavour we should not, and cannot, impose any sort of religious attitudes other than the ethical codes that prevail in our practice. But when it comes to the use of science on a practical level, and when it comes to utilising our discoveries in comprehending the value and the meaning of our existence, then the matter becomes completely different. This creates a framework in which both religion and faith play a role, and where an Islamic or non-Islamic attitude makes a difference.

To establish a worldview, we need to introduce our beliefs through direct and indirect connections. This is the role of culture in science. As explained above, culture and belief influence one's attitude and understanding of science. The richer the source of belief, the more influential it becomes in constructing the worldview of a nation or a faith. As Naquib al-Attas stated, "No true worldview can come into focus when a grand scale ontological system to project it is denied, and when there is a separation between truth and reality and truth and values."[15] Conversely, a worldview cannot be established without a well-constructed epistemology to assess the values of the practical and theoretical outcome of the scientific work. To say that the outcome of our modern scientific work is value-free is highly debatable, as argued by Aldemaro Romero Jr. in a

---

15 Sayed Muhammad Naquib al-Attas, Prolegomena to the Metaphysics of Islam: an exposition of the fundamental elements of the worldview of Islam, Malaysia Penerbit UTM Press, 2014, p.5.

recent article addressing the idea of predestination in biospeleology[16](a branch of biology dedicated to the study of organisms that live in caves).

The influence of an Islamic worldview on science is large, and an Islamic perspective can be formulated and presented on nearly every science topic. However, such perspectives are not singular; different scholars often form different attitudes. The differences in most cases are mainly in the details, as generally there is a common thread which unites these varying attitudes. All scholars affirm a belief in one Creator who is absolute, omniscient and omnipotent, and they all express a belief in predestination.

The Qur'an provides us with the basis to construct our worldview and shape attitudes towards our discoveries. It may help us see facts of nature that might otherwise be overlooked in the denial of value-free science— that is to say, science which does not recognise purpose and destiny. This would by no means denote the authority of religion over science; rather it would be an avenue through which we may elevate our understanding of science inasmuch as science can elevate our understanding of religion. To explain this, I will give two examples.

The first example is the Copenhagen interpretation of quantum mechanics, which allows for a particle to be present in two places at the same time. It is usually said that an electron can be found here in the lab and elsewhere in the world simultaneously. This conclusion is based on the fact that the solutions of the equations of quantum mechanics provide a probability distribution for finding the electron in space, and since the probability of finding an electron in any place is only part of the 100%, it is therefore claimed that this means that the electron is distributed all over the world in proportional values.

This bizarre interpretation can be discounted if we say that the electron is under continuous re-creation and is occupying different states through time, each of which has a definite probability allowed for by the

---

16 Aldemaro Romero Jr., "The influence of religion on science: the case of the idea of predestination in biospeleology", Research Ideas and Outcomes 2: e9015. 2016. DOI: 10.3897/rio.2.e9015.

solutions of the equations. Once we make a measurement of its position, we fix it to a position. So, although the theory is allowing the electron to be in different positions, it can actually only exist in one position. However, in any experiment, if the electron is shown to be in a superposition of more than one definite place, this is because the measurement time is too long compared to the re-creation time. Many other problems concerning the problem of quantum measurement can be solved by the assumption of re-creation which I have borrowed from the Islamic tradition of kalam.

The second example is the claim that the universe will continue to expand forever. This was concluded when cosmologists found that the universe is spatially flat and has a specific value for the average matter density. But according to the Qur'anic verse [21:104], the universe will eventually collapse. Such a contradiction can be resolved when we take into consideration that the same cosmological observations which tell us that the universe is flat, also tell us that it has a non-zero cosmological parameter. If we plug in a time-dependent cosmological parameter in the Einstein field equations for a flat spacetime and solve them, we find that there are solutions which allow for a collapsing universe. In fact we get a cyclic universe leading into a collapse phase, followed by a bounce when a new universe is created. This work has been verified by one of my postgraduate students.[17]

These two examples show that it is quite possible to have different interpretations for the same phenomena simply by adopting different epistemologies.

About the Scientific Miracles of the Qur'an

Over the past few decades, some writers have claimed that the Qur'an contains accurate scientific knowledge and that some scientific discoveries

---

[17] Munir Daradka, *A Collapsing Flat Universe*, M.Sc. thesis under the supervision of Prof. M.B. Altaie, Yarmouk University, Jordan (2007).

of the modern age were actually alluded to in the Qur'an over 1400 years ago. In most cases, these writers are unqualified to address this topic and the trend has evoked criticism from several points of view. One criticism is based on the view that the Qur'an is the divine word of God that will not change whilst science is susceptible to change. Therefore if we adopt one interpretation of a certain verse of the Qur'an today, we may have to change our interpretation in the future if a new scientific discovery is found to contradict it. The second criticism derives from the Qur'an itself, which states that human beings cannot grasp the true meaning of its content since Allah is the only One who knows.

One can see that the presentation of the scientific signs of the Qur'an under the umbrella term of *I'jaz al-'ilmi* has been, in some ways, a misuse of the truth of the Qur'an and the word of Allah. This is because it deviates from the original and sacred aim of revelation into the realm of claims that could lead to the illusion that the Qur'an is a book of science.

For Muslims, the Qur'an is the final word of the Divine, but this should by no means deprive us of the right to conclude some evidences of its authenticity as a divine source. To say that the verses of the Qur'an carry a 'multiplicity of layers of meanings' (an idea which is usually attributed to Ibn Rushd) is an inaccurate description of its content. It may create a fanciful, subjective understanding of the Qur'an which contradicts the Qur'an's statement in several verses that it has been revealed in clear Arabic. The correct statement is that the *wording* of the Qur'an allows for a 'multiplicity of layers of understanding'. Such layers are complementary to each other and not contradictory.

Once again I emphasise my stance that the Qur'an is not a book of science, but we cannot ignore the fact that its verses contain signs pointing to factual scientific meanings and descriptions. This is what makes the Qur'an a matchless divine text.

Does the Qur'an present Sacred Science?

Although the Qur'an is not a book of science, it presents a great deal of signs containing accurate knowledge which could contribute to an enlightened comprehension of the world. Such enlightenment helped the theologians of Islam in the past to recognise, as early as the second century after the Prophet Muhammad ﷺ, several basic principles and rules which established an original worldview and provided them with a profound Islamic approach to natural philosophy.

Harvard scholar Harry Wolfson expressed his astonishment at how the illiterate people of the Arabian peninsula could grasp so much knowledge in such a short amount of time.[18] They followed a methodological approach for deducing principles of thought that was distinct from that adopted by the Greek philosophers. William Craig has expressed this idea by saying, "The main difference between a Mutakallimūn (practitioner of kalam) and a Failasuf (philosopher) lies in the methodological approach to the object of their study: while the practitioner of kalam takes the truth of Islam as his starting-point, the man of philosophy, though he may take pleasure in the rediscovery of Qur'anic principles, does not make them his starting-point, but follows a method of research independent of dogma, without, however, rejecting the dogma or ignoring it in its sources."[19]

Through investigating the legacy of kalam, I found that the mutakallimūn devised five basic principles concerning the natural world, which counter the opposing doctrines adopted by the philosophers:[20]

(1) Temporality of the world (*Hudūth*): According to the mutakallimūn, the world is not eternal, but was created at some finite point in the past. Space and time had neither meaning nor existence before the creation of the world. Despite the fact that some of the mutakallimūn

---

18 H. A. Wolfson, *The Philosophy of the Kalam*, Harvard University Press, 1976.
19 W. L. Craig, *The Kalam Cosmological Argument*, p. 17.
20 Basil Altaie, *Daqīq al-Kalām*: the Islamic approach to Natural Philosophy, KRM, 2019.

believed that the original creation took place out of a pre-existing form of matter, the dominant view was that the creation took place *ex nihilo*—that is to say, out of nothing. Accordingly, they considered every constituent part of the world to be temporal.

(2) Discreteness (Atomism): The mutakallimūn believed that all entities in the world are composed of a finite number of fundamental elements, each called jawhar (the essence, substrate or substance), which is indivisible and has no parts. The jawhar was thought to be an abstract entity that acquires its physical existence, properties and value when occupied by a character called *'arad* (accident). These accidents are ever-changing qualities. Discreteness applies not only to material bodies but to space, time, motion, energy (heat), and all other properties of matter. Since the jawhar cannot stand on its own, as it would then be unidentified without being associated with at least one ʿarad, it can therefore be considered an abstract entity and a basic character that is different from the Greek and Indian atom. Some authors have tried in vain to relate the Islamic concept of the atom with those of the Greeks or the Indians. However, rigorous investigations have shown that it is unlikely that the Muslims took this idea from elsewhere; the Islamic atom possesses genuinely different properties.

(3) Re-creation and the ever-changing world: As Allah is the absolute Creator of the world and because He is living and always acting to sustain the universe, the mutakallimūn envisaged that the world would have to be re-created in every moment. This re-creation occurs with the accidents, not with the substances, but since the substances cannot be realised without being attached to accidents, the re-creation of the accidents effectively governs the ontological status of the substances too. Within such a process God stands to be the sustainer of the world. This principle is very important for two reasons: first, it establishes an indeterminate world; second, it finds a resonance in contemporary quantum physics.

(4) Indeterminism: Since Allah possesses absolute free will and since He is the Creator and the Sustainer of the world, He is then at liberty to take any action He wishes with respect to the state of the world or its control. Consequently, the laws of nature that we recognise have to be probabilistic and not deterministic so that physical values are contingent and undetermined. From such a theological stance, the mutakallimūn deduced the indeterminacy of events in the world. This resulted in denying the existence of deterministic causality, asserted by the fact that nature has no will of its own. The mutakallimūn also rejected the Greek's notion of the four basic elements and the alleged existence of any kind of intrinsic efficacy of those elements. This is one very central argument in kalam for the proof of the need for God; if nature is blind, no productive development can be expected.

(5) Integrity of space and time: The mutakallimūn had the understanding that space has no meaning on its own. Without there being a body, we cannot realise the existence of space. So is the case with time; it cannot be realised without the existence of motion, which needs a body to be affected. Imam al-Ghazali added a novel idea by suggesting that space and time should be treated on an equal footing. He negated the absoluteness of temporal allocations like the after and before and considered them relative to a certain reference point in time.

The above five principles have sound value in contemporary philosophy of science as they are in conformity with known facts about the natural world. The principles align with discoveries of modern physics and can be utilised to interpret many events and phenomena in the world. In this respect, the five principles work together to form one complete system through which phenomena of the world can be analysed and understood.

The signs that are given in the Qur'an provide measures and standards for the ethical treatment of many aspects of scientific research. For example, on the issue of environmental care, the Qur'an explains that much of the corruption that happens in nature is a result of the bad conduct of people and that humans will suffer the consequences of their corruption.

> *Corruption has appeared in the land and the sea on account of that which men's hands have wrought, that He may make them taste a part of that which they have done, so that they may return. [30:41]*

{ظَهَرَ الْفَسَادُ فِي الْبَرِّ وَالْبَحْرِ بِمَا كَسَبَتْ أَيْدِي النَّاسِ لِيُذِيقَهُمْ بَعْضَ الَّذِي عَمِلُوا لَعَلَّهُمْ يَرْجِعُونَ} [الروم:41]

Another example is the command to not change the creation of living animals genetically. This is implicitly contained in the verse from Surah al-Nisa':

> *I will order them to slit the ears of cattle, and indeed I will order them to change the nature created by Allah. [4:119]*

{وَلَأُضِلَّنَّهُمْ وَلَأُمَنِّيَنَّهُمْ وَلَآمُرَنَّهُمْ فَلَيُبَتِّكُنَّ آذَانَ الْأَنْعَامِ وَلَآمُرَنَّهُمْ فَلَيُغَيِّرُنَّ خَلْقَ اللهِ} [النساء:119]

The Islamization of Knowledge

In this context comes the question of the Islamization of knowledge, a project which aims to reform the presentation of knowledge in all specializations from an Islamic perspective. This is indeed an honourable objective; however, without having a strong philosophical base upon which such a project should stand, it is difficult to see how it can be executed and how the desired aims can be achieved.

The advocates of this endeavour claim that modern knowledge in the social, political, and natural sciences is cast in a Western mould based on materialistic perspectives and serves interests which are far removed from Islamic tenets. Their goal was announced in the First Conference on Muslim Education held in Mecca in 1977, where it was said that, "The task before us is to recast the whole legacy of human knowledge from the standpoint of Islam. In concrete terms, we must Islamize the disciplines in accordance with the Islamic vision."[21] This announcement did not gain much momentum despite the great efforts made by the conveners, namely, Syed Muhammad Naquib al-Attas (1931–) and Ismael al-Faruqi (1921–86), who were very enthusiastic about the project. Later, the International Institute of Islamic Thought (IIIT) was established to implement such aims, but with the exception of the publication of some books, very little was achieved and no clear working methodology was formulated.

In my opinion, the reason for the failure of the Islamization of knowledge proposal was its lack of philosophical basis. Beautiful dreams and passionate aims, although necessary, cannot guarantee success in such cases. The Islamization proposal did not have the profound vision by which intellectuals can be nucleated to work on a genuinely innovative programme that achieves its aims. This is why the conference was no more than a debate between conflicting views which could not find common ground, except in the aim of revitalizing the role of the Muslim nation (*ummah*) within contemporary international civilisation.

The system of knowledge is like a tree; we see the colourful and beautiful symmetrical upper part but we do not pay much attention to the lower parts, i.e. the roots that are necessary for the sustainment of the tree's life. Likewise, some Muslims look at the teachings of Islam and the social impact that it could have on the lives of individuals and societies, and they become eager to implement these ideals without understanding the prerequisites that must precede such outcomes.

Renowned characters such as Professor Abdus Salam, Seyyed Hossein Nasr and Ziauddin Sardar objected, at least partially, to the concept and the suggested methodology of the Islamization of knowledge. I would agree with the late Professor Abdus Salam in identifying science as being an outcome which is independent of any religion or belief. It is the implementation of science and its application which may be tinged with special religious and philosophical beliefs. However, when we analyse

---

21 al-Attas, p. 7.

the philosophical basis of certain scientific assumptions, we may find that there is a kind of embedded character within those assumptions that subscribes to certain beliefs. This might be clear in social sciences, as they involve many assumptions and analyses which depend on underlying philosophies and views; nonetheless, one can find examples from natural sciences too. For example, to assume that the universe is eternal embeds (at face value) the assumption that there is no creator or that God is the universe itself.

This is the reason why Stephen Hawking questioned a place for the Creator once he discovered that the universe is non-singular (has no beginning in time). Whereas, in fact, this result does not negate the necessity of the Creator once we realise that creation and re-creation is happening all the time. In another example, to assume that the laws of nature work independently would be to eliminate divine action, except perhaps for coordinating the action of such laws. Nevertheless, a theistic science need not involve divine action in every step of natural processes; it suffices to assume that necessity is at play with the available possibilities and that the outcome of the action of natural laws is indeterministic—implying necessarily the need for an agency not belonging to the same set of laws—to determine the results.

To summarise, I would say that introducing Islamic tenets into science can be made possible if an underlying philosophy is found to furnish a background for the views. In this case, daqiq al-kalam is best suited to play such a role. Problems in natural as well as social sciences and arts can be analysed, studied, and interpreted in the light of its principles. Accordingly, a wealth of knowledge that has a common and consistent basis might be formed, which can then constitute a coherent body of knowledge that reflects an Islamic contribution to civilisation.

Science is not an ideology and should not be set in ideological moulds of any form, whether Marxist, Christian, Islamic or otherwise. History of science and empirical experience tells us that scientific knowledge should satisfy four basic requirements to be called 'scientific': it should be explanatory, consistent, verifiable (or falsifiable), and predictive. Knowledge that does not explain anything or is in self-contradiction, or is not verifiable, cannot be called scientific knowledge. This is the fundamental difference between science and faith. On the other hand, predictability is a strong element in any theory as it can be used as a direct means of verification. Accordingly, the 'Islamization of knowledge' should contribute to the meaningful content of scientific knowledge and enhance one or more of the elements in its structure; otherwise, it is a useless exercise sought in vain. This might be a challenging goal to achieve; nevertheless, it is an acceptable one. Through the contributions of religion, we can help enhance the value of science as well as make science enhance our understanding of religion. Otherwise, religion should remain an independent authority exercising its authority in matters of faith alone.

Islam deals with knowledge in a general sense. It does not provide us with sacred science but with sacred illumination, to seek our destiny by observing both the seen and the unseen. The Qur'an makes it clear that life does not end through death but extends much beyond our life in this world. Accordingly, and despite the fact that we cannot prove the existence of the afterlife physically, the Qur'an helps us to find evidences regarding the purpose of our life as well as a glance at scientific views where a realisation of some matters of faith can be made possible.

Modern physics, specifically relativity theory and quantum mechanics, continues to provide us with knowledge that opens up a new realm in dealing with consciousness and the meaning of reality. There is now ample opportunity to explore these topics and use such knowledge to construct physical possibilities for topics which were once considered to be limited to faith. Indeed, the consciousness of human beings can never be destroyed, as I will explore later. The more we grasp of science and understand the world, the more we approach nearer to the truth,

The Qur'an and Science

We will show them Our Signs in the universe, and in their own selves, until it becomes manifest to them that this (the Qur'an) is the truth. Is it not sufficient in regard to your Lord that He is a Witness over all things? [41:53]

{سَنُرِيهِمْ آيَاتِنَا فِي الْآفَاقِ وَفِي أَنْفُسِهِمْ حَتَّى يَتَبَيَّنَ لَهُمْ أَنَّهُ الْحَقُّ أَوَلَمْ يَكْفِ بِرَبِّكَ أَنَّهُ عَلَى كُلِّ شَيْءٍ شَهِيدٌ} [فصلت:53]

Chapter Three

The Solar System

The solar system is composed of the Sun, planets, satellites, asteroids and comets. The most distant planet belonging to this system is around 40 times the distance between the Earth and the Sun (about 150 million kilometers or 93 million miles), whereas the farthest edge of the solar system is around 100 times this distance.

Formation of the Solar System

It is now agreed among astrophysicists that the solar system was formed approximately 5 billion years ago from the condensation of a gaseous cloud, the remnants of a supernova explosion. The supernovae are exploding stars inside which huge fusions of chemical elements result in the formation heavier of elements. What remains of such explosions is a huge gaseous nebula. The gases, rubble and dust of such nebulae gather gradually over a long period of time around the densest part of the nebula, forming a nucleus of attraction. Massive amounts of the nebula then condense to form a star. At the same time, and in other parts of the nebula, other centres of density might also be formed where much smaller amounts of matter gather to form large stones roughly 10 km in size called (planetesimals). These arrange in rings around the central part—the star under formation—and gradually collect through mutual gravity collapsing on the denser point in the region to form a planet.

During the past three decades, astronomers have discovered hundreds of planetary systems belonging to other stars. These are called exoplanets. Most of the discovered planets are large gaseous planets and a minority are small. The large size of the majority of the exoplanets may be caused by detection techniques which have limited sensitivity. Increasing the sensitivity of the detection may help in the discovery of smaller planets.

In some cases, the formed planets are not solid but remain as gaseous giants. This applies to the four planets in our solar system: Jupiter, Saturn, Uranus and Neptune. Each planet then carries its motion around the newly-formed Sun in nearly circular orbit. Other objects such as planetesimals may remain or may get ejected by the gravity of the planets to far distances and form a belt of icy objects surrounding the solar system. Such objects, known as comets, are attracted by the Sun and frequently visit it, appearing as stars with tails when coming nearby.

There are some important points to consider here:
1. The solar system has formed out of a huge cloud of smoke and dust.
2. The formation of the solar system, including the Sun and the planets, has taken a significant amount of time.

These points are important for understanding the content of the next section which deals with the formation of the solar system as presented in the Qur'an.

The Sun

The Sun is a nearly perfect sphere composed mainly of hot highly ionized gases known as plasma. It is by far the most important source of energy for life on earth. Its diameter is about 1.4 million kilometers (109 times that of earth), and its mass is around 330,000 times that of the Earth, which means it makes up approximately 99.86% of the total mass of the solar system. About three quarters of the mass of the Sun is composed of hydrogen (~76%) and the rest is mostly helium (~24%), as well as much smaller amounts of the heavier chemical elements, including carbon, oxygen, neon, and iron.

According to astrophysicists, our sun is thought to have been formed about 4.6 billion years ago out of a nebula (a huge gaseous cloud) which was the remnant of a previously existing massive star that ended its life in a supernova explosion, thus ejecting a huge amount of heavy element-rich dust and gases.

The surface temperature of the Sun is roughly 6000 degrees Kelvin and its core temperature exceeds 15 million degrees Kelvin. It revolves on an axis perpendicular to the plane of rotation in a period of about one month, although this rotation is not uniform; the polar parts rotate faster.

The Sun ejects an intense stream of electric charges moving at high speed, accompanied by strong magnetic fields generated by the fast motion of these charges. These are called solar winds, sometimes forming strong storms. Should these storms hit the Earth, they can cause considerable damage. The magnetic field of the Earth prevents such charges from reaching its surface, shielding itself from immense damage.

The Sun also emits a large amount of short wavelength radiation which is very harmful to living tissues. The Earth is saved from this harmful radiation by its atmosphere; specifically, by the ozone layer.

The Sun in the Qur'an

The Qur'an mentions the Sun 33 times, as a sight that we see during the day. No metaphoric meaning is given here. The Sun is part of the creation of Allah and is specifically designated as being an intrinsically shining body, emitting light as a candle, unlike the Moon which is a shining body that reflects light. It is also said that the Sun and the Moon are moving in their due course with accurate calculable timing.

The Divine Word and the Grand Design

> *The Sun and the Moon run on their fixed courses (exactly) calculated with measured out stages for each. [55:5]*

{الشَّمْسُ وَالْقَمَرُ بِحُسْبَانٍ} [الرحمن:5]

Indeed, this is alluding to the fact that people use the apparent motion of the Sun and the Moon to calculate time and fix the calendar.

The Fate of the Sun

Much emphasis has been placed in the Qur'an on the relationship between doomsday and the fate of the Sun. In this respect, there are several verses regarding the development phases of the Sun until the end of time. In Surah al-Taqwir the Qur'an reveals that the Sun will be dimmed in a major collapse during which time the stars will look fuzzy—that is, at the end of time:

> *When the Sun is folded up. And when the stars are dust-coloured [81:1–2]*

{إِذَا الشَّمْسُ كُوِّرَتْ (1) وَإِذَا النُّجُومُ انْكَدَرَتْ } [التكوير:1-2]

For a long time, people could not understand the source of the immense energy radiated by the Sun. In 1934, Hans Bethe was able to explain how the Sun produces such an amount of energy employing the nuclear fusion of hydrogen and converting it into helium, the second most abundant element after the former. This means that billions of megaton hydrogen bombs are exploding inside the Sun every second. It was further discovered that the process inside the Sun and the other stars was some kind of nuclear 'cooking' process by which light elements are converted into heavier ones, releasing large amounts of energy. This process basically converts a small fraction of the Sun's mass into energy. Billions and billions of hydrogen atoms get converted into helium every second, generating huge amounts of energy which get radiated in the form of heat, light and other radiation. According to the estimated amount of hydrogen available for nuclear fusion in the Sun, astrophysicists have calculated that the Sun can maintain its brightness for another five billion years.

The Sun is an automated thermodynamic system. The nuclear fusion is maintained at an almost steady rate through the balance between the gravitational forces acting inward which contract the body of the Sun, and the internal outward pressure produced by nuclear fusion causing the Sun to expand. Once the gravitational forces act to contract the body of the Sun into a smaller size, the pressure and temperature inside the core will increase in accordance with the laws of thermodynamics. This in turn will automatically enhance the rate of nuclear reactions, causing an increase in the internal pressure and making the core expand.

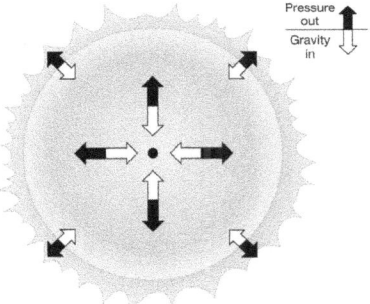

Fig. I The Sun

As the core expands, it becomes cooler and the rate of the nuclear reactions drops, reducing the pressure inside the core and allowing gravity to play the role of contracting it again. This interplay between gravitational forces and nuclear reactions is the secret behind the balanced stable solar furnace. The Sun is said to be sustained by hydrostatic equilibrium. Each second, four million tons of mass is converted into heat and radiation.

After billions of years, the amount of hydrogen inside the core of the Sun will be reduced to a level insufficient to maintain the rate of nuclear fusion necessary to keep it stable, while the amount of helium in the core will increase. As the rate of fusion lowers, the pressure inside the core decreases substantially, causing the core to contract.

Subsequently, the pressure and the temperature will rise tremendously until it reaches the required level where helium atoms are fused to form carbon. Three helium atoms are fused to form one carbon atom. Such a process releases a great amount of energy. What is remarkable is that the process of helium fusion occurs suddenly in all parts of the core in an explosive manner; unlike the process of hydrogen fusion, which takes place at a controlled, steady rate.

The Red-Giant

The colossal explosion of the core causes the Sun to expand suddenly and become over 100 times larger than its original size. As this happens, the outer part of the Sun becomes red because the expansion reduces its surface temperature, and at this stage the Sun is said to become a red-giant; a monster. Some astrophysicists claim that the Sun will then swallow the planets Mercury and Venus and become very close to the Earth's surface. Others propose that it may swallow the Earth too. In any case, for a hypothetical observer watching this dramatic event, the sky will appear as if it is torn and reddish. The Qur'an describes this event:

So when the heaven is rent asunder, so it becomes red like red hide.
[55:37]

{فَإِذَا انْشَقَّتِ السَّمَاءُ فَكَانَتْ وَرْدَةً كَالدِّهَانِ} [الرحمن:37]

In another verse, the Qur'an states:

And the heaven will split asunder, for that day it will be frail.
[69:16]

{وَانْشَقَّتِ السَّمَاءُ فَهِيَ يَوْمَئِذٍ وَاهِيَةٌ} [الحاقة:16]

As the Sun nears the Earth, the atmosphere of the Earth will disappear (the heat will make it evaporate) and consequently the Earth will be naked. We read in Surah al-Taqwir in the same context:

> *And when the heaven shall be stripped off and taken away from its place. [81·11]*

{وَإِذَا السَّمَاءُ كُشِطَتْ} [التكوير:11]

Indeed, this is a very accurate description of what could happen to the Earth's atmosphere. Note here that the meaning of the word 'heaven' is associated with the atmosphere.

The Final Stage: The White-Dwarf

The Sun may remain at the stage of a red-giant for thousands of years and some hydrogen in the outer parts of its core may also fuse again, but the amount of hydrogen available for such a fusion will soon be exhausted. Eventually, the whole body of the Sun will collapse colossally to become a small, very hot but very faint object called the white-dwarf. The size of the Sun will be smaller than the present size of the Earth, with a diameter of only around 10,000 km. Since its core is mainly composed of compressed carbon, it would appear like a huge diamond in the sky.

The process of the final collapse of the Sun under the gravitational pull of its own parts is described in the Qur'an as *taqwir*. According to the Arabic lexicon of Ibn Faris,[22] this word means to clump an object onto itself by evolving it together, like the saying 'to clump the turban'. In exactly the same manner, the Sun will go through its final stage of gravitational collapse, clumping its matter onto itself while revolving around its axis.

The Qur'an does not mention that the Earth will be joined together with the Sun but has explicitly stated that the Sun and the Moon will be joined together:

---

22 Ibn Faris, *Mu'jam al-Maqayis Fi al-Lugha*, 2nd Edition, Dar al-Fikr, Beirut, 1998.

The Divine Word and the Grand Design

And the Sun and Moon will be joined together. [75:9]

{وَجُمِعَ الشَّمْسُ وَالْقَمَرُ} [القيامة:9]

This may indicate that when the Sun becomes a red-giant, the Moon will be at its nearest distance from the Sun. This occurs when it is in conjunction with the Sun on the last day of the lunar month.

This leads to another point that may require explanation. For a long time, some Muslims were confused about the meaning of a verse in Surah Ya'sin [36:38]. This confusion is reflected in different translations of this verse as follows:

Khan: *And the Sun runs on its fixed course for a term (appointed). That is the Decree of the All-Mighty, the All-Knowing.*

Maulana: *And the Sun moves on to its destination. That is the ordinance of the Mighty, the Knower.*

Pickthall: *And the Sun runneth on unto a resting-place for him. That is the measuring of the Mighty, the Wise.*

Sarwar: *How the Sun moves in its orbit and this is the decree of the Majestic and All-Knowing God.*

Shakir: *And the Sun runs on to a term appointed for it; that is the ordinance of the Mighty, the Knowing.*

Sherali: *And the Sun is moving on to its determined goal. That is the decree of the Almighty, the All-Knowing God.*

Yusuf Ali: *And the Sun runs his course for a period determined for him: that is the decree of (Him), the Exalted in Might, the All-Knowing.*

Clearly, there are significant differences in the above translations. However, it is also clear that some translators have recognised that the verse is pointing to a determined goal or destiny for the Sun, but do not specify this destiny with a specific position as other translators have indicated.

In my opinion, the running of the Sun is not to be understood as moving in space—although it is doing so—but that it is evolving through different stages to reach its final destiny. Accordingly, an accurate translation of the verse may read:

> *And the Sun moves on to its destiny of stability. That is the ordinance of the Mighty, the Knower. [36:38]*

{وَالشَّمْسُ تَجْرِي لِمُسْتَقَرٍّ لَهَا ذَلِكَ تَقْدِيرُ الْعَزِيزِ الْعَلِيمِ} [يس:38]

In this case, we identify the Sun's destiny with the last stage of the *stable* white-dwarf. The reason I emphasise the word stable is because not all white-dwarfs are so.

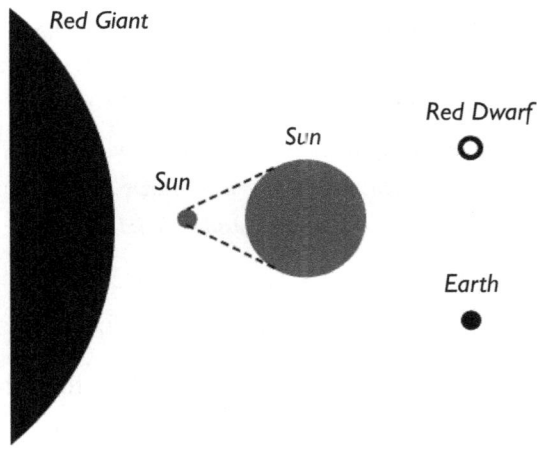

Fig. 2 The Sun compared at different stages of its life

Sun Rising from the West?

It might be worth discussing two cultural narrations of the Prophet Muhammad ﷺ regarding this future prediction. The first is a *hadith* which describes how the Sun will be brought very near to the Earth on doomsday. This is already expected to happen in the above scenario sug-

The Divine Word and the Grand Design

gested by modern astrophysics, whereby the Sun will be very close to the Earth, if not to swallow it.

The second hadith narrates that one of the signs of doomsday is the Sun rising from the West. How could this be? Does the rotation of the Earth reverse, and if so, why?

If we return to the above scenario describing the astrophysical fate of the Sun, we notice that the formation of the red-giant will happen suddenly and the huge sudden expansion of the Sun may take only a few hours as a result of the explosive fusion of the helium nuclei inside the core. If such an event is to happen after sunset (for a particular observer) then the Sun will re-appear from the West due to its enormous size filling a substantial part of the Western sky at once. This situation on doomsday is pointed out in the Qur'an in several verses:

> *When the heaven is split asunder. And listens and obeys its Lord, and it must do so. And when the Earth is stretched forth. It will eject its contents, as it erupts. [83:1–4]*

{إِذَا السَّمَاءُ انْشَقَّتْ (1) وَأَذِنَتْ لِرَبِّهَا وَحُقَّتْ (2) وَإِذَا الْأَرْضُ مُدَّتْ (3) وَأَلْقَتْ مَا فِيهَا وَتَخَلَّتْ} [الانشقاق:1-4]

Here we notice the statement that the sky will be torn, the Earth will expand and its interior will be emptied. Such a scene is well-anticipated from an astrophysical point of view as the Sun becomes a red-giant approaching very close to the Earth. In fact, the Earth is expected to melt, and this is what is meant by the Earth being 'stretched forth'.

Yet, in another verse the Qur'an states:

> *When the heaven is Cracked. And when the planets are scattered. And the oceans are exploded. [82:1–3]*

{إِذَا السَّمَاءُ انْفَطَرَتْ (1) وَإِذَا الْكَوَاكِبُ انْتَثَرَتْ (2) وَإِذَا الْبِحَارُ فُجِّرَتْ} [الانفطار:1-3]

The Solar System

In this verse, even more detail is given: the sky becomes cracked, the planets are scattered and the oceans explode. Again, such a scene is expected once the red-giant comes very near to the Earth. The magnetosphere and the atmosphere protecting the Earth will be cracked and a protecting shield will no longer be available. In the same section, the Qur'an presents a similar image:

> *The Day that the sky will be like the boiling filth of oil. And the mountains will be like flakes of wool. [70:8–9]*

{يَوْمَ تَكُونُ السَّمَاءُ كَالْمُهْلِ (8) وَتَكُونُ الْجِبَالُ كَالْعِهْنِ} [المعارج:9-8]

This is again a very dreadful scene that can be pictured in light of the Sun becoming a red-giant and approaching the Earth. The flares of such a gigantic burning body will make the sky appear like a boiling fluid of molten stones, metals and burning gases.

Speculative Calculations

In the following section, I will present some speculative calculations concerning a time when the whole universe may enter a state of collapse, on the premise that the universe will indeed collapse as stated in verse [21:104].

As understood by previous commentators, if we take the verse in which the unreferenced 50,000 years is mentioned [70:4] as the duration for the collapse of the universe, and if the duration of a year is taken as 365.2422 days, then this will result in exactly 18,262,110 days. If we then take each day to be 1000 years of what the Arabs were counting, then this means that the duration of the collapse of the universe (duration of doomsday according to commentators) will be 18,262,110,000 lunar years. This is around 17.718 billion tropical years. According to the most recent estimations, the age of the universe now is 13.772 billion tropical years. This means that the universe will start collapsing in just less than 4 billion years from now. Such an estimation is not very

The Divine Word and the Grand Design

far from the estimated duration remaining for the Sun to exhaust its efficient amount of fuel and collapse and become a red-giant, and eventually a white-dwarf.

According to this calculation, I conclude that by the time the Sun goes into collapse (taqwīr), the whole universe will be collapsing into a big crunch. However, such a collapse will take about 17 billion years to be completed. This calculation is based on interpretation of the Holy Book and should not be taken to establish any fact other than an indication that doomsday will commence when the Sun is completely exhausted and the final taqwīr takes place. In other words, the above calculations affirm that the collapse of the Sun marks the beginning of doomsday as described in the Qur'an.

The Moon

The Moon is the only known satellite for the Earth and has a mass of less than 1/80 of the Earth's. It rotates around the Earth in 29.5 days in a nearly circular orbit. The reason for this difference is the motion of the Earth around the Sun.

The Moon also revolves around an axis passing through its centre in 29.5 days and this is the reason why it appears the same every month over the years, without change. The distance between the Earth and the Moon is about 384,000 km and its diameter is about a quarter of the diameter of the Earth.

The Moon has no atmosphere and for this reason, its surface is full of craters produced by falling meteorites. It has no magnetic field due to the absence of a molten core. The surface of the Moon is covered with very fine dust mostly produced by the falling celestial debris.

Gravity on the Moon is less than one-fifth of that on the surface of the Earth. Objects on the surface of the Moon weigh less and for this reason, astronauts are seen jumping rather than walking on the Moon. The American astronaut who first landed on the Moon in 1969 was filmed jumping. If one is to walk with ease on the Moon, it would be necessary to wear heavy shoes made of lead.

The Moon is a frequent object of admiration for poets and dreamers. Many poems across all languages are written regarding the beauty of the Moon and the thoughts and feelings inspired upon viewing it.

Fig. 3 The Moon

The Moon in the Qur'an

The Moon has been mentioned in the Qur'an 27 times, mostly in conjunction with mention of the Sun. The Qur'an tells us that the Sun and the Moon were created to be subservient to humans. This is indeed very clear in the case of the Sun, but how can it be that the Moon is subservient to us? One possible answer is that we need the Moon as a reference for calculating time, as the lunar month is a natural period by which we can reckon the passage of time and the lunar calendar is the oldest type of calendar. Apart from this, it has recently been discovered that the presence of the Moon is necessary for developing life on this planet. The tidal effect of the Moon generates the movement of water in large lakes and seas, thus producing effects on ponds in areas near lakes

The Divine Word and the Grand Design

and seas. Such ponds are a necessary environment for the development of primitive life.[23]

The angular size of the Moon is equivalent to the angular size of the Sun, which allows for the total eclipse of the Sun at times. If the angular size of Moon was smaller, no total eclipse would occur.

The Sun's light is reflected from the surface of the Moon and this is how we see its surface. The Qur'an specifies this by saying that the Moon is made to shine whereas the Sun is made to generate light like a candle. This is indeed the case, as the Sun generates heat and light whilst the Moon reflects the light received from the Sun.

> *He it is Who made the Sun a shining brightness, and the Moon a light, and ordained for it stages that you might know the computation of years and the reckoning. Allah created not this but with truth. He makes the signs manifest for a people who know. [10:5]*

{هُوَ الَّذِي جَعَلَ الشَّمْسَ ضِيَاءً وَالْقَمَرَ نُورًا وَقَدَّرَهُ مَنَازِلَ لِتَعْلَمُوا عَدَدَ السِّنِينَ وَالْحِسَابَ مَا خَلَقَ اللَّهُ ذَلِكَ إِلَّا بِالْحَقِّ يُفَصِّلُ الْآيَاتِ لِقَوْمٍ يَعْلَمُونَ}[يونس:5]

Note that this verse emphasises the creation of the Sun and the Moon with 'truth', which alludes to the law of nature (or the Divine sunnah) by which such events and periods run.

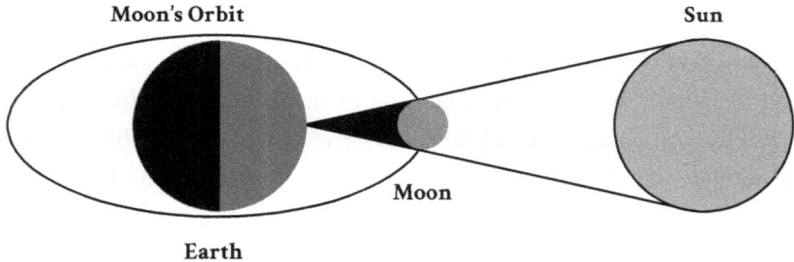

Fig. 4 The Sun and the Moon in the last ay of the month

---

23 For details, see: Bruce Dorminey, 'Without the Moon, would there be a life on Earth?' *Scientific American*, April 21, 2009.

One face of the Moon

We normally see the same face of the Moon. We never see the back of the Moon, despite the fact that the Moon is rotating around an axis passing through its centre. Why is this?

The reason is that the period of the rotation of the Moon around its axis is equal to the period of its rotation around the Earth, so that when the Moon makes a quarter of a revolution around the Earth, it makes a quarter of a revolution around itself and so on. As such, the part of the Moon facing the Earth maintains the same position as seen from the Earth, shown in the figure below.

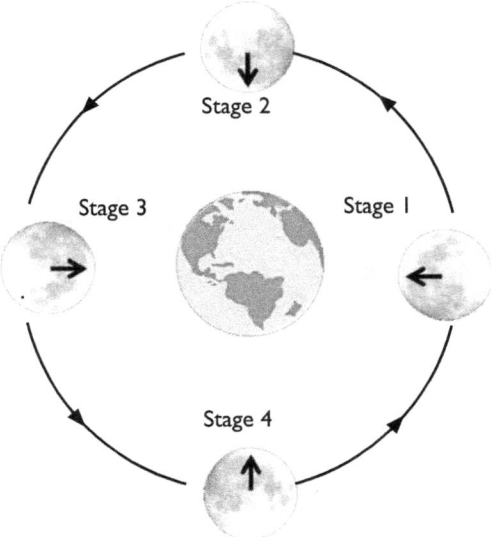

Fig. 5 Locked rotation of the Moon.

This is also the case for some planets outside of our solar system. On these planets, the days and nights are eternal. Should this happen in the rotation of the Earth around the Sun, then half the globe would be in perpetual daytime and the other half in perpetual night-time.

The Divine Word and the Grand Design

The Moon appears to be moving in the constellations and this motion has been mentioned in the Qur'an:

> He it is Who made the Sun a shining brightness, and the Moon a light, and ordained for it *manāzil* (allocated positions in the zodiac) that you might know the computation of years and the reckoning. Allah created not this but with truth. He makes the signs manifest for a people who know. [10:5]

{هُوَ الَّذِي جَعَلَ الشَّمْسَ ضِيَاءً وَالْقَمَرَ نُورًا وَقَدَّرَهُ مَنَازِلَ لِتَعْلَمُوا عَدَدَ السِّنِينَ وَالْحِسَابَ مَا خَلَقَ اللَّهُ ذَلِكَ إِلَّا بِالْحَقِّ يُفَصِّلُ الْآيَاتِ لِقَوْمٍ يَعْلَمُونَ} [يونس:5]

In another verse, the Qur'an mentions the phases and their relation to the *manāzil*, which are allocated positions on the zodiac:

> And the Moon, We have ordained for it stages till it becomes again like an old dry palm-branch. [36:39]

{وَالْقَمَرَ قَدَّرْنَاهُ مَنَازِلَ حَتَّى عَادَ كَالْعُرْجُونِ الْقَدِيمِ} [يس:39]

That is, the Moon becomes feeble and withered as it approaches the last few days of the lunar month.

Finally, the Qur'an also tells us that the Sun and the Moon will be joined together on doomsday:

> When the eye is dazzled. And the Moon is cracked. And the Sun and Moon are joined together.[24] [75:7–9]

{فَإِذَا بَرِقَ الْبَصَرُ (7) وَخَسَفَ الْقَمَرُ (8) وَجُمِعَ الشَّمْسُ وَالْقَمَرُ} [القيامة:7–9]

This is quite possible as the Sun will become a red-giant extended in size to reach the Earth's orbit. Furthermore another verse in the Qur'an states that as doomsday nears, the Moon will rupture. This is already

---

[24] Some authors have mistakenly translated the word *khasafa* which means cracked as being eclipsed. This is not the case since that case it should have been written *khusifa*.

anticipated, as the tidal force and the heat of the red-giant may cause such an effect.

Note that in the above verse the wording used to describe the Moon as being 'cracked' is very precise, although some readers may interpret the above verse as suggesting that the Moon will be eclipsed. This is not the case, however; as explained in the footnote, the word does indeed mean cracked. The Sun will swallow the Moon while it is eclipsed by the Earth unless it swallows the Earth first, since at the time of the lunar eclipse, the Earth is situated between the Sun and the Moon with the latter being at the far side.

The Hour of Doom is drawing near, and the Moon is rent asunder.[25] *[54:1]*

{اقْتَرَبَتِ السَّاعَةُ وَانْشَقَّ الْقَمَرُ} [القمر:1]

If so, and if the Moon is to be swallowed exclusively by the Sun and not the Earth, then such an event is expected to happen during the last day of a lunar month when the Moon is nearest to the Sun.

The Earth

The Earth is the most beautiful known planet, not only in our solar system, but in the whole universe. It contains large areas of green life and the majority of its surface is covered with huge amounts of water. Its complex gaseous atmosphere and magnetic shield protects it from the fatal dangers of the sky. This kind of construction is by no means an act of mere chance, since the construction of the Earth involves the very accurate coordination of several elements to make life possible.

The Earth is a small planet amongst eight others rotating around the Sun at different distances. Measured with reference to the distance of the Earth from the Sun in AU (astronomical unit, equivalent to 149.6 million

---

[25] This verse is claimed to have been revealed to the Prophet on the occasion of a miracle by which the Moon was visually split into parts.

kilometers), the following list outlines the distance from the planets to the Sun in ascending order: Mercury (0.4 AU far from the Sun); Venus (0.7 AU); Earth (1.0 AU); Mars (1.5 AU); Jupiter, the largest planet (5.2 AU); Saturn (9.5 AU); Uranus (19.2 AU); Neptune (30.0 AU) and finally, the dwarf planet Pluto (39.5 AU). None of these planets, with the exception of the Earth, is known to harbour life.

The Earth rotates around the Sun in an elliptic orbit with an average distance of 149.6 million kilometers, covering a full period in 365.2411 days. The Earth spins around an axis passing through its centre, completing one full period in 24 hours. With this motion, we see the Sun rising daily from the East and setting in the West. Such periodic movement allows for the day and night to follow each other in succession. The axis of spin is inclined by 23.5 degrees from the line perpendicular to the plane of rotation around the Sun. This inclination generates tropical seasons on its surface.

The mass of the Earth is about 6×10^{24} kilograms (trillion billion metric tons). This mass allows it to have adequate gravity on its surface and helps produce suitable bodies for animals and human beings, allowing easy mobility. Obviously, this does not imply that the mass of the Earth is tailored to suit the needs of its inhabitants; it could be the other way around. The Earth's inhabitants could have been developed to suit the Earth's surface gravity and environment.

Biology tells us that liquid water is the most essential element for life in the universe. The search for life on other planets, including those hosted by stars other than the Sun (called exoplanets), puts the availability of liquid water on the planet's surface as the most basic requirement for habitability. Accordingly, the term 'habitable zone' has been coined, to mean the zone around the star in which liquid water can exist on the surface of the planet. In our solar system, the habitable zone is between Venus and Mars.

The Earth in the Qur'an

The Earth has been mentioned in the Qur'an 451 times, often in conjunction with the Sun. Here it is important to note that in Arabic the word 'Earth' bears several meanings. The Qur'an mentions the word Earth (*ard* أرض) with three meanings, which are:

1. The limited piece of land.

Do you not see that Allah sends down water from the cloud so the Earth becomes green? [22:63]

{أَلَمْ تَرَ أَنَّ اللَّهَ أَنْزَلَ مِنَ السَّمَاءِ مَاءً فَتُصْبِحُ الْأَرْضُ مُخْضَرَّةً} [الحج:63]

2. The village, town, or state.

Verily, Pharaoh behaved arrogantly in the Earth. [28:4]

{إِنَّ فِرْعَوْنَ عَلَا فِي الْأَرْضِ} [القصص:4]

3. The planet.

Verily! In the creation of the heavens and the Earth, and in the alternation of night and day, there are indeed signs for men of understanding [3:190]

{إِنَّ فِي خَلْقِ السَّمَاوَاتِ وَالْأَرْضِ وَاخْتِلَافِ اللَّيْلِ وَالنَّهَارِ} [آل عمران:190]

The different meanings are dependent on the context of the verse. A reader of Arabic can guess the intended meaning of the word, but a commentator should take extra care when considering the meaning by studying the context of the verse. This is crucial to avoid misinterpretation and the erroneous conclusion that there is a contradiction between the Qur'an and our modern knowledge.

Evidence for a Revolving Earth

There are several indications in the Qur'an of the rotation of the Earth around its axis. The main evidence is the expression given to describe the succession of day and night.

The Divine Word and the Grand Design

> *He has created the heavens and the Earth with truth. He **yukawirū** the night the day and **yukawirū** the day on the night. [39:5]*

{خَلَقَ السَّمَاوَاتِ وَالْأَرْضَ بِالْحَقِّ يُكَوِّرُ اللَّيْلَ عَلَى النَّهَارِ وَيُكَوِّرُ النَّهَارَ عَلَى اللَّيْلِ} [الزمر:5]

In the verse it is said that Allah yukawirū the night over the day and yukawirū the day over the night. This Arabic word means that night (darkness) comes over the day (light) in rotational motion and that the day (light) comes over the night (darkness) in rotational motion. Arabic lexicons give the analogy of a turban where the cloth is collected over its parts in rotational motion. So here we can assert that this expression, which is rarely used in Arabic, indicates the rotation of the Earth around its axis.

Another Qur'anic verse in this respect says:

> *And He placed stabilizers (mountains) on Earth, lest it tumbles with you. [16:15]*

{وَأَلْقَى فِي الْأَرْضِ رَوَاسِيَ أَنْ تَمِيدَ بِكُمْ} [النحل:15]

Other similar verses are [21:31] and [31:10]. The word *tamīd* is similar to *tamīl* (lean) with one important difference. The word tamīd means tamīl through motion. This is the reason for not using the word tamīl in this verse; it embeds an implicit meaning suggesting that if the solid body of the Earth was unbalanced, it would tumble. This would indeed be expected to happen if the Earth became unbalanced. This is yet another example of the precise and careful choice of words in the Qur'an.

Water and Life

Water is very important for life since it has specific properties that are necessary for living systems to work. Water has a relatively wide range of temperatures through which it stays in the liquid state, between 0 and 100 degrees Celsius. It is a very good solvent due to the polarity of its molecules. This property is necessary for the mobility of materials through the body of living things. Without water, solid materials cannot be fused to the cells in a living body and no growth can be expected. The Qur'an says:

> We made from water every living thing. Will they not then believe? [21:30]

{وَجَعَلْنَا مِنَ الْمَاءِ كُلَّ شَيْءٍ حَيٍّ أَفَلَا يُؤْمِنُونَ} [الأنبياء:30]

Water covers a large part of the Earth's surface. Scientists believe that part of this water has accumulated from the condensation of vapour arising out of volcanoes. However, it seems that this is not the sole source of water on the Earth. Investigating the history of the Earth's development led scientists to discover that large amounts of water are derived from sources away from the Earth. According to the research work of Alessandro Morbidelli,[26] most of today's water originates from proto-planets formed in the outer asteroid belt that plunged towards Earth. This was supported by the measured deuterium/hydrogen proportions in carbon-rich chondrites. The water in carbon-rich chondrites points to a similar deuterium/hydrogen ratio in oceanic water. The Qur'an states:

> And We send down water from the sky in precise measure, then We caused it to settle in the Earth, and We indeed are able to carry it away. [23:18]

{وَأَنزَلْنَا مِنَ السَّمَاءِ مَاءً بِقَدَرٍ فَأَسْكَنَّاهُ فِي الْأَرْضِ وَإِنَّا عَلَى ذَهَابٍ بِهِ لَقَادِرُونَ} [المؤمنون:18]

This is notable for being the only verse out of many verses regarding the fall of water from the sky, in which the Qur'an specifies the fallen water being sent in a measure (*bi qadar*) and made to reside (*askannāhu*) in the Earth. This may point to the water residing in the oceans, as in this context we take the word 'Earth' to mean the whole planet. However, it could also be understood to mean underground reservoirs as well.

---

26 Alessandro Morbidelli et al., *Meteorites & Planetary Science* 35, 2000, 1309–1329

Flatness of the Earth

Some have claimed that the Qur'an tells us that the Earth is flat. However, one should be careful as the word Earth (ard) carries more than one meaning; it could mean the region of land, as for example in the verse [2:17], but it could also mean the whole planet as in the verse [11:7] and many other verses.

In this context, we can understand the verses [12:3], [48:48] and [71:19]. However, there is a verse which says:

> And after that He spread (**dahāhā**) the Earth. [79:30]

{وَالْأَرْضَ بَعْدَ ذَٰلِكَ دَحَاهَا} [النازعات:30]

We understand from Arabic etymology that the word *dahāhā* means leveling by artificial or natural means. The next verse mentions allowing water to spread the land and allow the plants to grow. This may point to the natural means by which limited regions of land were leveled. However, the question remains as to why the Qur'an has broached this topic with expressions that are unclear. The answer is that this topic was one that could not be explored explicitly at the time of revelation. If the Qur'an had stated that the Earth is round, it would have given strength to the argument of the Prophet Muhammad's ﷺ opponents, who accused him of being mad. However, there are other verses implicitly pointing to a spherical Earth and its spin, as shown in the following verse:

> He created the heavens and the Earth truthfully. He rolls the night over the day, and rolls the day over the night. [39:5]

{خَلَقَ السَّمَاوَاتِ وَالْأَرْضَ بِالْحَقِّ يُكَوِّرُ اللَّيْلَ عَلَى النَّهَارِ وَيُكَوِّرُ النَّهَارَ عَلَى اللَّيْلِ} [الزمر:5]

Here, the word *yukawiru* has been translated as 'rolls' and this is the accurate literal meaning of the word, free from the influence of the translator's beliefs. If the Earth is flat, why would the Qur'an use such a delicate description as the word yūkawiru? Moreover, the Qur'an has indicated that the alteration of the day and night is a matter for contemplation and thinking (See [2:164], [3:190], [10:6] and [45:5]). This suggests that the inspiration for thinking about how day and night follow each other only makes sense if the Earth is spherical and is rotating on an axis.

The Qur'an presents the ontological argument for the Creator and His creation by directing the attention of the readers to clear markers that they see in their environment, i.e. the land around them, the sky above them and various aspects of their life. This provides simple and direct evidence to support the claim for one Creator. Besides this, the Qur'an includes some subtle hints to inspire those who think and contemplate on their belief in the Divine source. This explains why people accepted the argument for the truth of the Prophet Muhammad ﷺ and the truth of his narration. Obviously some of them could not accept the claims made by the Qur'an and presented counterarguments, but once we analyse these arguments, we can see that they were mainly arguing against the abstractness of the Creator and what they perceived to be the irrationality of the promised afterlife. Their minds could not comprehend such metaphors. This was the reason behind their insistence on being presented with miraculous evidence in support of the claims made by the Prophet Muhammad ﷺ. This is similar to an incident in which the atheist particle physicist Steven Weinberg challenged his theist colleague John Polkinghorne during a public debate, asking for a fiery sword to appear from nowhere and hit him for his impiety. This chimes with what the unbelievers asked of the Prophet Muhammad ﷺ in order to prove his truthfulness:

So cause a piece of the heaven to fall on us, if you are of the truthful! [26:187]

{فَأَسْقِطْ عَلَيْنَا كِسَفًا مِنَ السَّمَاءِ إِنْ كُنْتَ مِنَ الصَّادِقِينَ} [الشعراء:187]

In a similar argument, the physicist Lawrence Krauss argued that he would be ready to believe in God if one night he woke up to see that the stars in the sky aligned to read 'I am here'. Such arguments against the belief in one Creator are based on the presumption that God rules the universe with miracles, rather than order. On the contrary, the Qur'an states that Allah has created the heavens and the Earth with truthfulness and such truthfulness can only be achieved if there is a law by which we can evaluate the occurrence of events and their due course in the world. If we can recognise that the world is organised and well-ordered, this gives evidence for the presence of a creator and sustainer of the world. To have a world being subject to law and order does not necessarily imply its intrinsic sovereignty unless we can prove that every component of the world has its own decision-making authority contributing to related events. No such luxury is available in running our universe. Laws of nature need an operator since they are indeterministic, while laws of physics are our own mental constructions.

The Atmosphere

Would it be possible for planet Earth to harbour life without having such a delicate construction and position in space? The Earth enjoys the presence of an essential mixture of gases surrounding it and extending tens of kilometres into space. This mixture of gases is carefully provided to enable life to flourish on this planet and without such a careful arrangement of the atmosphere, no developed life on Earth could be expected.

The atmosphere is composed mainly of the following gases: 78% nitrogen, an inert gas essential for the biochemical compounds; 21% oxygen, which is essential for oxidation and growth of living creatures; and water vapour, which accounts for roughly 0.25% of the atmosphere by mass. Other constituents are argon gas and carbon dioxide with traces of helium, hydrogen, methane and krypton making up the remainder. This gaseous mixture maintains its composition through complicated processes of depletion and compensation.

In addition to being essential for the ingredients of life, the gaseous sphere surrounding the Earth provides necessary protection to living creatures.

The atmosphere provides the necessary cover to maintain a limited range of temperature variations during day and night. If the amount of carbon dioxide increased by a small fraction then the heat emitted from the Sun would become trapped and the temperature of the Earth would rise tremendously. This would result in runaway global warming that would lead to the extinguishment of life on Earth, as is the case on Venus.

The atmosphere is subdivided into several layers and according to the most famous description, these are:

- Troposphere: 0 to 12 km
- Stratosphere: 12 to 50 km
- Mesosphere: 50 to 80 km
- Thermosphere: 80 to 700 km
- Exosphere: 700 to 10,000 km

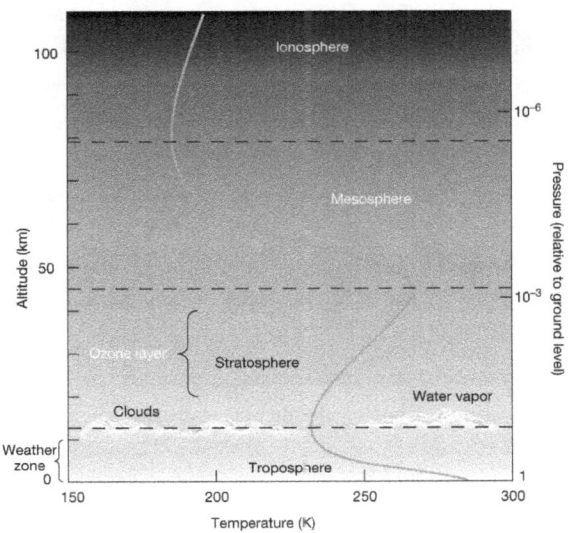

Fig 6 The Earth's atmosphere

Temperature varies in these layers, decreasing through the troposphere and increasing through the stratosphere then decreasing through the rest of layers.

The density of air decreases as we go up as well as the amount of oxygen. This fact is reflected in the Qur'an: *Allah will open the hearts of whomever He wants to guide to Islam, but He will tighten the chest of one whom He has led astray, as though he was climbing high up into the sky.* [6:125]

The atmosphere is a very thin layer. The Kármán line, at 100 km (which makes about 1.57% of Earth's radius) is used as the border between the atmosphere and outer space. This allows one to consider the atmosphere as a thin skin surrounding the Earth.

The atmosphere provides the Earth with vital protection from the following dangers:

1. Meteorites falling daily from the sky with different sizes and mass. If these small stones reached the Earth's surface, no living creature on land would survive. Most of the falling debris from space becomes hot as it passes through the atmosphere and melts and evaporates before reaching the surface of the Earth.
2. Short wavelength radiation like ultraviolet radiation and the gamma rays are absorbed by the ozone layer surrounding the Earth. Such radiation could cause harmful damage to the skin as well as a higher evaporation rate from the oceans, and could leave the surface of the earth without any green area, like Mercury.
3. The atmosphere causes the scattering of the Sun's light, allowing for light to be dispersed through space and helps to illuminate places that are not exposed to direct sunlight.

The Clouds

Clouds have been mentioned in the Qur'an 9 times. In one verse, the Qur'an beautifully and precisely describes the formation of the clouds:

> See you not that Allah drives the clouds gently, then joins them together, then makes them into a heap of layers, and you see the rain comes forth from between them. And He sends down from the sky hail (like) mountains, (or there are in the heaven mountains of hail from where He sends down hail), and strike there with whom He will, and averts it from whom He wills. The vivid flash of its (clouds) lightning nearly blinds the sight. [24:43]

{أَلَمْ تَرَ أَنَّ اللَّهَ يُزْجِي سَحَابًا ثُمَّ يُؤَلِّفُ بَيْنَهُ ثُمَّ يَجْعَلُهُ رُكَامًا فَتَرَى الْوَدْقَ يَخْرُجُ مِنْ خِلَالِهِ وَيُنَزِّلُ مِنَ السَّمَاءِ مِنْ جِبَالٍ فِيهَا مِنْ بَرَدٍ فَيُصِيبُ بِهِ مَنْ يَشَاءُ وَيَصْرِفُهُ عَنْ مَنْ يَشَاءُ يَكَادُ سَنَا بَرْقِهِ يَذْهَبُ بِالْأَبْصَارِ} [النور:43]

Indeed, this description of the formation of clouds and the nucleation of the drops of water with some turning into frozen droplets is a magnificent presentation of details that could not be known to people of the desert during the time of revelation. This verse is very compelling in consideration of the Qur'an as a revealed text from a divine authority.

The verses which mention the clouds and their usefulness are also mentioned in the context of the effect of the wind. It is notable that the Qur'an draws the attention of the reader to consider the wind and the clouds as part of the signs that prove the existence of the Creator and highlight the glory of creation.

> In the creation of the heavens and the Earth, and in the alternation of night and day, and the ships which sail through the sea with that which is of use to mankind, and the water (rain) which Allah sends down from the sky and makes the Earth alive therewith after its death, and the moving (living) creatures of all kinds that He has scattered therein, and in the veering of winds and clouds which

The Divine Word and the Grand Design

are held between the sky and the Earth, are indeed Ayat (proofs, evidences, signs, etc.) for people of understanding. [2:164]

{إِنَّ فِي خَلْقِ السَّمَاوَاتِ وَالْأَرْضِ وَاخْتِلَافِ اللَّيْلِ وَالنَّهَارِ وَالْفُلْكِ الَّتِي تَجْرِي فِي الْبَحْرِ بِمَا يَنْفَعُ النَّاسَ وَمَا أَنْزَلَ اللَّهُ مِنَ السَّمَاءِ مِنْ مَاءٍ فَأَحْيَا بِهِ الْأَرْضَ بَعْدَ مَوْتِهَا وَبَثَّ فِيهَا مِنْ كُلِّ دَابَّةٍ وَتَصْرِيفِ الرِّيَاحِ وَالسَّحَابِ الْمُسَخَّرِ بَيْنَ السَّمَاءِ وَالْأَرْضِ لَآيَاتٍ لِقَوْمٍ يَعْقِلُونَ} [البقرة:164]

In this, the Qur'an is presents part of the ontological argument which is frequently used to present the glory of the creation.

Meteorites

A meteorite is a solid piece of debris from a comet, asteroid, or meteoroid that originates in outer space and survives its passage through the atmosphere to reach the surface of the Earth. When such an object enters the atmosphere, various factors such as friction, pressure, and chemical interactions with atmospheric gases cause it to heat up and radiate energy in the form of light. It then becomes a meteor—a glowing streak of light—and may form a fireball. Such objects are also known as shooting stars or falling stars. Thousands of meteorites enter the atmosphere everyday but only a few reach the Earth's surface. Meteorites vary greatly in size and only a few of them can produce a creator. The surface of the Moon is full of creators because it has no atmosphere.

A few times a year, the orbit of the Earth passes through a region that contains debris from passing comets. This debris is formed of small particles, stones and dust. As the Earth passes through a cloud of debris, these particles fall on the Earth and a firework-like show takes place during certain nights where hundreds of thousands of meteors are seen in the sky. These are called meteor-showers.

The Solar System

Fig. 7 Meteor shower

The Qur'an has mentioned meteorites in several places and has described them as stony pieces falling from the heavens. This is clear in a verse in which the opponents of the Prophet Muhammad ﷺ challenged his claim that the Qur'an is the truth revealed from Allah: *They also say, "Lord, if this (Qur'an) is the Truth from you, shower down stones on us from the sky (instead of rain) or send us a painful punishment."* [8:32]. Also:

> *Let parts of the sky fall on us if what you say is true.* [27:187]

{فَأَسْقِطْ عَلَيْنَا كِسَفًا مِنَ السَّمَاءِ إِنْ كُنْتَ مِنَ الصَّادِقِينَ} [الشعراء:187]

In this verse, the 'parts of the sky' (kisaf) are stones.

Furthermore, the Qur'an has criticised unbelievers at the time of revelation for confusing meteorites with atmospheric phenomena like heaped clouds.

> *And if they were to see a piece of the heaven falling down, they would say: "Clouds gathered in heaps!"* [52:44]

{وَإِنْ يَرَوْا كِسْفًا مِنَ السَّمَاءِ سَاقِطًا يَقُولُوا سَحَابٌ مَرْكُومٌ} [الطور:44]

Indeed, according to the Aristotelian vision meteors were considered to be an atmospheric phenomenon. Here the Qur'an dismisses such an understanding by saying that the observed meteors are material pieces (stones) falling from the sky. This is an important piece of information which was certainly unknown to the Prophet Muhammad ﷺ at the time of revelation and is a sign of the divine source of the Qur'an.

The Magnetic Shield

The Earth has a magnetic field generated by rotating molten core which is composed mainly of ionized atoms of iron and nickel. The motion of charged atoms produces a magnetic field according to the laws of electromagnetism. This makes the Earth behave like a bar magnet, inclined by about 11 degrees with respect to the Earth's rotational axis. This produces a magnetic sphere surrounding the globe with its north pole located near the geographical North Pole and the south magnetic pole located near Greenland.

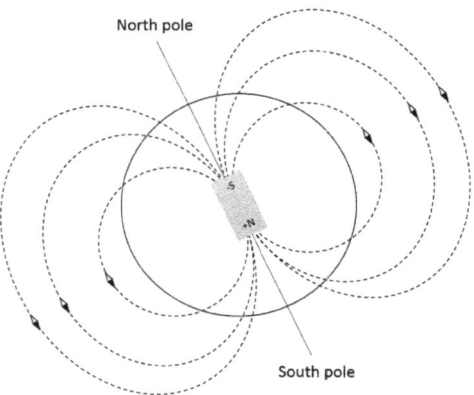

Fig. 8 Earth's magnetic field

The magnetic field of the Earth is weak and extends for tens of thousands of kilometres into space, but it is essential for life-providing protection against the high-speed electrically-charged particles coming with the solar wind bombarding the Earth. The magnetic field deflects these charged particles and directs them into high altitude above the ionosphere. Accordingly, two belt-shaped reservoirs of charged particles are formed at high altitude and these are called Van Allen Belts, named after their discoverer. The two main belts extend from an altitude of about 500 to 58,000 kilometres.

As the charged particles of the solar wind are deflected at high altitudes, a natural colourful light display appears in the Earth's sky, predominantly seen in the high-latitude regions around the Arctic and Antarctic. This is called an aurora and is sometimes referred to as polar lights, northern lights (aurora borealis) or southern lights (aurora australis).

The positions of the magnetic poles vary widely over long-time scales, but sufficiently slowly for ordinary compasses to remain useful for navigation. However, at irregular intervals averaging several hundred thousand years, the Earth's field reverses and the North and South Poles abruptly switch places. This reversal of the geomagnetic poles leaves a record in rocks that are of value to scientists in calculating geomagnetic fields of the past. In turn, such information is helpful in studying the motions of continents and ocean floors in the process of plate tectonics.

Without the magnetic field, the ozone layer—which protects the Earth from harmful short wavelength radiation— and the upper atmosphere would be stripped off by solar winds and cosmic rays.

The role of the Earth's magnetic field is part of an overall role performed by its atmosphere. This helps to explain the meaning of the verse: *And We have made the sky well-guarded roof, yet they turn away from its signs* (i.e. Sun, Moon, winds, clouds, etc.). The Qur'an has summed up the role of the atmosphere and the magnetosphere in providing necessary protection to life on Earth:

> *And We have made the heaven a guarded canopy; yet they turn away from its signs. [21:32]*

{وَجَعَلْنَا السَّمَاءَ سَقْفًا مَحْفُوظًا وَهُمْ عَنْ آيَاتِهَا مُعْرِضُونَ} [الأنبياء:32]

The Divine Word and the Grand Design

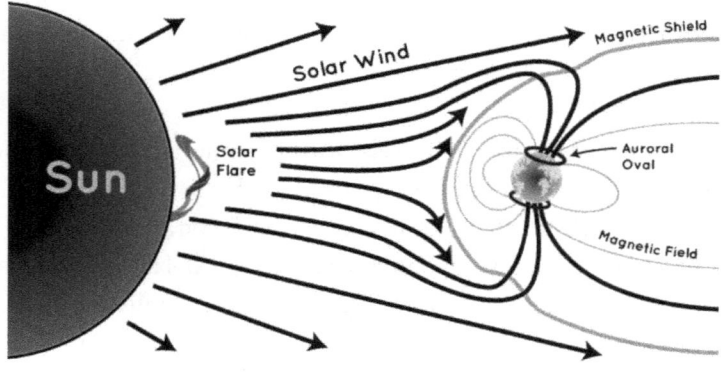

Fig. 9 The magnetic shield

Indeed, the heavens act like a guarded shield protecting the creatures on Earth's surface, otherwise life on the planet would be impossible. Other planets like Mercury and Mars do not enjoy such a luxury. Meteorites fall on Mercury in large numbers everyday and the solar wind storms the surface of Mars without protection.

The Day and the Night

The word 'day' has been mentioned in the Qur'an 57 times. The word 'night' has been mentioned 92 times. It is mentioned in the Qur'an that the day covers the night:

> *He covereth the night with the day, which is in haste to follow it, and hath made the Sun and the Moon and the stars subservient by His command. [7:54]*

{يُغْشِي اللَّيْلَ النَّهَارَ يَطْلُبُهُ حَثِيثًا وَالشَّمْسَ وَالْقَمَرَ وَالنُّجُومَ مُسَخَّرَاتٍ بِأَمْرِهِ أَلَا لَهُ الْخَلْقُ وَالْأَمْرُ تَبَارَكَ اللَّهُ رَبُّ الْعَالَمِينَ} [الأعراف:54]

Hence, it is the day which covers the night and the day follows the night hastily. This could also be suggestive of the spinning of the Earth. It is also interesting to note that the Qur'an describes the day covering the night like a veil.[27] This indicates that the day is only regional and covers a limited area surrounding the Earth. This is indeed the case, since the day is manifested through the luminous semi-sphere covering half of the Earth's body at any time. This is elaborated in the following verse:

> *And a sign for them is the night from which WE peel off the day, and they are left in darkness. [38:37]*

{وَآيَةٌ لَهُمُ اللَّيْلُ نَسْلَخُ مِنْهُ النَّهَارَ فَإِذَا هُم مُّظْلِمُونَ} [يس: 37]

Why would the Qur'an use the word 'peel' (*naslakhu*) in this context? The answer lies in the fact that the day is demonstrated by the atmosphere, which scatters the light spreading in all directions. This enables us to see things without being exposed to direct sunlight. On the Moon, which has no atmosphere, light is not scattered and we cannot see things unless exposed to direct sunlight. For this reason, the shining observable part of an object will be that part which receives light directly from the source. On Earth we see objects under the table because light is scattered by the atmosphere. The Earth's atmosphere is a very thin layer of about 100 km. Compared to the radius of the Earth (which is about 6000 km) this is only 1.6% of the size of the Earth. Such a thickness is comparable to the thickness of the skin of a cow, for example. When the day is removed by the setting of the Sun, it is peeled off the Earth as peeling off the skin of a cow.

We should note here that the Qur'an does not state that the night is peeled off the day; it is the day which is peeled off the night. The reason for this is that the night (darkness) is the basic character of the universe, not the day (illumination). The sky looks black on the Moon and the shade is a completely dark area, unlike on the surface of the Earth where the shade is not as dark since the atmosphere scatters light to allow some illumination. For this reason, the word 'night' always precedes the word 'day' in the Qur'an. What subtlety and greatness lies in the Qur'an!

---

27 The word *ughshi* means covering with a thin veil. See verse [10:27].

The Divine Word and the Grand Design

Fig. 10 The shade is very dark on the moon.

Going to Outer Space

The Qur'an has alluded to the possibility of man going to outer space by saying that this is only possible with certain means and power:

> *Jinn and mankind, if you can escape through the diameters of the heavens and the Earth, do so, but you cannot do so without power and authority. [55:33]*

{يَا مَعْشَرَ الْجِنِّ وَالْإِنسِ إِنِ اسْتَطَعْتُمْ أَن تَنفُذُوا مِنْ أَقْطَارِ السَّمَاوَاتِ وَالْأَرْضِ فَانفُذُوا ۚ لَا تَنفُذُونَ إِلَّا بِسُلْطَانٍ} [الرحمن:33]

The Qur'an has also pointed to the fact that outer space is totally dark:

> *And even if We open to them a gate of heaven, and they keep on ascending through it. They would say: Oh! It is as if our eyes have been covered over, or rather we are bewitched. [15:14–15]*

{وَلَوْ فَتَحْنَا عَلَيْهِم بَابًا مِّنَ السَّمَاءِ فَظَلُّوا فِيهِ يَعْرُجُونَ (14) لَقَالُوا إِنَّمَا سُكِّرَتْ أَبْصَارُنَا بَلْ نَحْنُ قَوْمٌ مَّسْحُورُونَ} [الحجر:14-15]

The reason for this is that space is dark when astronauts pass through the effective region of the atmosphere where light is scattered. Past this region, they will suddenly feel that they are in the dark as if their eyes have been covered or they will feel enchanted. How did the Prophet Muhammad ﷺ know about this?

It is remarkable that whenever the Qur'an talks about motion in outer space the word *ya'ruju* and its derivatives are used. This word specifically designates ascendance along a curved trajectory. This is very accurate because motion in outer space always happens along curved geodesics as in the theory of general relativity (see the figure below). Even the light rays in space follow curved trajectories.

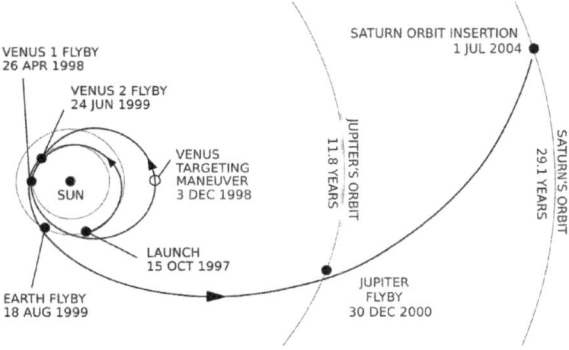

Fig. 11 *Trajectories of motion in space*

Many references to geodesic motion are made in the Qur'an:

He sends the order from the heavens to the Earth, and then the order ascends to Him in a day which is equal to one thousand years of yours. [32:5]

{يُدَبِّرُ الْأَمْرَ مِنَ السَّمَاءِ إِلَى الْأَرْضِ ثُمَّ يَعْرُجُ إِلَيْهِ فِي يَوْمٍ كَانَ مِقْدَارُهُ أَلْفَ سَنَةٍ مِمَّا تَعُدُّونَ} [السجدة:5]

It seems that even the angels follow geodesic paths in space rather than moving in straight lines: *To Him ascend (ta'ruju) the angels and the Spirit*

The Divine Word and the Grand Design

in a day the measure of which is fifty thousand years. [70:4] Also, in Surah Saba' we read:

> *He knows what goes into the Earth and what comes forth from it, what descends from the heaven and what ascends (ya'ruju) thereto.*
> *[34:2]*

{يَعْلَمُ مَا يَلِجُ فِي الْأَرْضِ وَمَا يَخْرُجُ مِنْهَا وَمَا يَنْزِلُ مِنَ السَّمَاءِ وَمَا يَعْرُجُ فِيهَا وَهُوَ الرَّحِيمُ الْغَفُورُ} [سبأ:2]

The same expression is used in a verse in Surah al-Hadid:

> *He knows what goes into the Earth and what comes forth from it, what descends from the heaven and what ascends (ya'ruju) thereto.*
> *[57:4]*

{يَعْلَمُ مَا يَلِجُ فِي الْأَرْضِ وَمَا يَخْرُجُ مِنْهَا وَمَا يَنْزِلُ مِنَ السَّمَاءِ وَمَا يَعْرُجُ فِيهَا} [الحديد:4]

Chapter Four

The Stars

Stars are huge celestial bodies like our Sun, formed mainly of the gases hydrogen and helium. Some stars are smaller than the Sun in mass and size, some are like the Sun, and there are others which are much larger. Stars were formed at the early stages of the developing universe within the early gaseous nebulae. The accumulation of gases at points with relatively high density causes the pressure, temperature and gravity to grow, gradually attracting even more gases. Consequently, the pressure and the temperature rise until reaching a level where hydrogen nuclei are diffused, forming helium atoms and releasing a large amount of energy in the form of light, heat and other radiation. At that moment, a fully-fledged star is born.

Astronomers have now observed the stages of star formation through sensors on board satellites dedicated for such purposes. They can even photograph stars during their early stages of formation. Once a star is formed and attains its stable state it will continue propagating heat, light and other radiation for a long time—typically several billions of years. A star's life depends mainly on its initial mass. The larger the mass, the shorter the life of the star, since for a larger mass the rate of nuclear fusion is higher, which means faster consumption of its material. Low mass stars live longer; some even live for 20 billion years or more.

Some stars are much larger than our Sun; especially those formed at the beginning of the developing universe. Within such stars, chemical elements heavier than hydrogen were formed through the processes of nuclear fusion. So, we can say that stars are gigantic nuclear power plants automatically controlled by the laws which are part of their design.

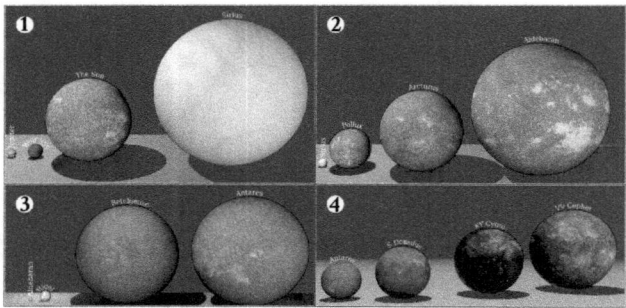

Fig. 12 Comparison of the sizes of the stars

Stars have different sizes, masses and temperatures. We can recognise the surface temperature of a star by analysing the light coming from it. A blue star is a hot star with a surface temperature of about 10,000 degrees and more, while a red star is relatively cold with a surface temperature of 2000–3000 degrees. Looking up into the night sky, you may be able to identify the famous Orion constellation, the hunter, shown in the figure below. At the top left there is one red star with the Arabic name Betelgeuse and there is another star toward the lower right with the Arabic name Rigel which means 'leg'. The constellation was labelled 'the hunter' by the ancients due to its shape. Betelgeuse is larger than our Sun and is more than 700 million times its size, whereas Rigel is 500,000 times larger than our Sun.

Stars get grouped in galaxies, with a typical galaxy containing 100-200 billion stars. The observable universe contains more than 300 billion galaxies.

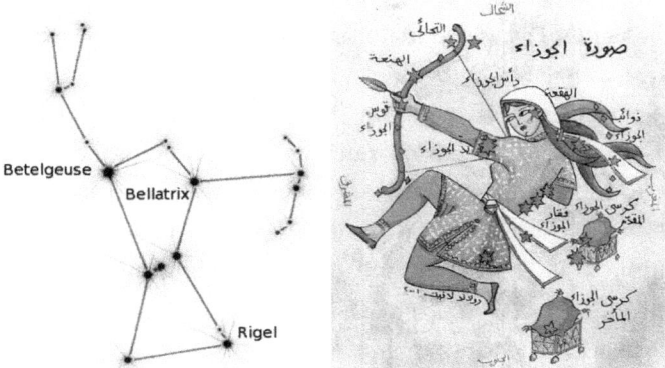

Fig. 13 The constellation Orion

The Fate of Stars

The active life of a star comes to an end when nuclear fusion ceases due to a deficient amount of nuclear fuel. Consequently, the pressure inside the star drops and the star collapses under its own gravity. Such a collapse continues until it reaches a stable state by which the internal forces inside the star balance the gravitational forces causing the collapse; otherwise, the rising temperature and pressure in the core of the star would cause the elements to fuse and form heavier elements. In the case of our Sun, the temperature and pressure produced by the collapse causes the helium nuclei to fuse in a process called helium flash, forming carbon. This second stage of fusion causes the body of the Sun to expand and become a red-giant, as explained earlier. Upon the collapse of the red-giant, in the final stage the Sun becomes a white-dwarf, a stable object smaller than our Earth with a high surface temperature of about 20,000 degrees, emitting white light and barely visible with the naked eye from a far distance.

Neutron Stars

If the star is of a larger initial mass, around 1.4–3.4 solar mass, then it will have a different fate. The burning of the star will cause the formation of heavier elements reaching iron. As the star passes through the stages, it will finally collapse and form a neutron star where all the atoms in the core get crushed and converted into neutrons. This is called a neutron star. It is roughly the size of a small city about 10 kilometers in radius, rotating about its axis rapidly and generating a huge magnetic field.

Supernova Explosion

Some of these stars are formed out of unstable white-dwarfs when a nearby companion is available which has reached the stage of a red-giant. The strong gravity of the white-dwarf causes the gases and materials of the red-giant to be sucked in by the white-dwarf and consequently a disk, called an accretion disk (composed of the compressed material) is formed around the white-dwarf. As the flow of material towards the white-dwarf continues, the temperature and the pressure rises tremendously and the rate of nuclear fusion increases very rapidly to form the elements heavier than iron up until the heaviest element, uranium. Such a formation will suddenly cause a huge explosion emitting a very large amount of energy in what is called a supernova explosion, which leaves behind a small, rapidly rotating, pulsating neutron star. This is called a pulsar.

Supernova explosions are accompanied by bursts of acoustic shock waves, like a knocking in the heavens.

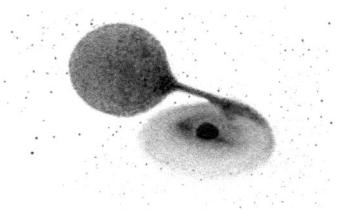

Fig. 14 The Giant and the Dwarf

Pulsars

Pulsars are objects that were discovered in 1969 when intense radio waves emitted from an unknown source with accurate pulse repetition frequency were detected. When first discovered, it was thought that the source was caused by intelligent beings living in that region but it was soon realised that the source was a rapidly rotating neutron star, the remnant of a supernova explosion.

The supernova explosion causes the sudden appearance of a flashing star in the sky with very high brightness. Such an event happened in 1054 AD and has been observed and documented by the Arabs and Chinese. Another similar event was observed by Galileo Galilei in 1604 AD.

Black Holes and Wormholes

Giant stars with a mass larger than 3.5 times the mass of the Sun will not stop at the stage of becoming a neutron star after their final collapse. Such stars continue their collapse and all the matter they contain crushes to a point. This point is surrounded by a sphere called the event horizon at which all events disappear and even time is frozen. Such an object cannot be observed since even light cannot escape the gravity of its event horizon. If our Sun is imagined to turn into a black hole, then the radius of its event horizon will be about 3 kilometers and no more.

The concept of black holes is very attractive to laymen, who are curious to know what lies beyond the event horizon. The fact is that even specialised physicists do not fully know. It is possible that it is an empty region as everything should collapse to the point at the centre, known as the singularity. The black hole could be a gateway to another world, or at least, this is what some theoretical physicists believe. Such a gate opens at the throat of a wormhole. The wormhole is a channel, going through which would cause one to appear in another place in the universe.

The Divine Word and the Grand Design

Stars in the Qur'an

Stars have been mentioned in the Qur'an 13 times. In one verse the Qur'an points to the use of stars in finding the 'ways'. In several other verses, the stars are considered subservient to human beings.

And He has subjected to you the night and the day, the Sun and the Moon; and the stars are subjected by His Command. Surely, in this are proofs for people who understand. [16:12]

{وَسَخَّرَ لَكُمُ اللَّيْلَ وَالنَّهَارَ وَالشَّمْسَ وَالْقَمَرَ وَالنُّجُومُ مُسَخَّرَاتٌ بِأَمْرِهِ إِنَّ فِي ذَلِكَ لَآيَاتٍ لِقَوْمٍ يَعْقِلُونَ} [النحل:12]

It is remarkable that the verse ends with the line "Surely, in this are proofs for people who understand"—a point to contemplate.

In another verse the Qur'an tells us that the allocation of the positions of the stars is something of majestic importance.

I swear by the positions of the stars. This is an oath, if you only knew, that is awesome. [56:75–76]

{فَلَا أُقْسِمُ بِمَوَاقِعِ النُّجُومِ (75) وَإِنَّهُ لَقَسَمٌ لَوْ تَعْلَمُونَ عَظِيمٌ} [الواقعة: 75–76]

Indeed, it is hard to know for certain the accurate positions of the stars since they are at large distances and affected by many factors, including their motions, which are observer-dependent. Many factors that go into defining the instantaneous positions of the stars involve some appreciable inaccuracies.

In several other verses, stars are mentioned in the context of doomsday and how they will appear at that time.

In one verse Allah swears by the heaven and the *Tariq*. He then describes the Tariq as a star which has piercing brightness:

By the heaven, and the Tariq. Do you know what the Tariq is? The star of piercing brightness. [86:1–3]

The Stars

{وَالسَّمَاءِ وَالطَّارِقِ (1) وَمَا أَدْرَاكَ مَا الطَّارِقُ (2) النَّجْمُ الثَّاقِبُ}
[الطارق: 1–3]

Certainly, this verse is describing an unusual star. In fact, Tariq in Arabic means 'the knocker'. Hence this verse incorporates, in very few words, the precise description of two features of a supernova explosion: the intense piercing brightness and the knocking which accompanies such an event. Such periodic knocking also occurs with the pulsar, as shown clearly by computer simulations of received signals.

Chapter Five

Time

Time is a very puzzling concept. Some scientists like Albert Einstein consider time to be a kind of illusion, whilst others think it is a real entity independent of any other thing, associated with an eternal flow of change.

There are two types of time: physical and psychological. The former is a measure of change, normally taken in terms of the periodic motion of astronomical objects like the Sun (defining the day), the Moon (defining the month) and the alteration of seasons (defining the tropical year). These are natural units for measuring time which have been divided further into smaller conventional units like hours, minutes and seconds.

The second type is psychological time which is measured in connection with personal feelings. This is the time generated by psycho-physiological effects. For example, a secretion of hormones may cause different effects in the body and may even cause changes in the activities of cells. Although modern science considers psychological time to exist only when the human being is in a conscious state, practical experience (such as sleeping) tells us that some awareness of psychological time exists even when we are unconscious.

The Divine Word and the Grand Design

Time in the Qur'an

Both types of time have been mentioned in the Qur'an; physical time is measured with reference to astronomical periods (year, month, day), and psychological time is estimated by people who are either unaware of physical time or those who return to consciousness after being in an unconscious state. For example:

> *Or like the one who passed by a town and it had tumbled over its roofs. He said: "Oh! How will Allah ever bring it to life after its death?" So Allah caused him to die for a hundred years, then raised him up (again). He said: "How long did you remain (dead)?" He (the man) said: "(Perhaps)* **I remained (dead)** *a day or part of a day." He said: "Nay, you have remained (dead) for a hundred years." [2:259]*

{أَوْ كَالَّذِي مَرَّ عَلَى قَرْيَةٍ وَهِيَ خَاوِيَةٌ عَلَى عُرُوشِهَا قَالَ أَنَّى يُحْيِي هَذِهِ اللَّهُ بَعْدَ مَوْتِهَا فَأَمَاتَهُ اللَّهُ مِائَةَ عَامٍ ثُمَّ بَعَثَهُ قَالَ كَمْ لَبِثْتَ قَالَ لَبِثْتُ يَوْمًا أَوْ بَعْضَ يَوْمٍ}
[البقرة:259]

This example is among several mentioned in the Qur'an indicating that psychological time during a state of unconsciousness exists, but can only be estimated once we return to consciousness.[28] This conclusion, once applied to other verses regarding the resurrection on doomsday, implies that consciousness is restored to bodies of the dead. For this reason, when the deceased are asked about how long they have remained dead, they will perceive it to be a very short time.

> *On the day when the Hour of Doom comes, the criminals will swear that they have remained (in their graves) for no more than an hour. They had been deceived in this way. Those who have given knowledge and have faith will say: By the decree of God, you have remained for the exact period which was mentioned in the Book of*

---

28 See also verse [18:19].

God about the Day of Resurrection. This is the Day of Resurrection, but you did not know. [30:55–56]

{وَيَوْمَ تَقُومُ السَّاعَةُ يُقْسِمُ الْمُجْرِمُونَ مَا لَبِثُوا غَيْرَ سَاعَةٍ كَذَلِكَ كَانُوا يُؤْفَكُونَ (55) وَقَالَ الَّذِينَ أُوتُوا الْعِلْمَ وَالْإِيمَانَ لَقَدْ لَبِثْتُمْ فِي كِتَابِ اللَّهِ إِلَى يَوْمِ الْبَعْثِ فَهَذَا يَوْمُ الْبَعْثِ وَلَكِنَّكُمْ كُنْتُمْ لَا تَعْلَمُونَ} [الروم:55–56]

This indicates that the dead do not feel a sense of time and also that some kind of consciousness returns to the body upon resurrection. However, resurrection may not mean returning to the same physical status of the body as during life on Earth, since it has been mentioned in the Qur'an that resurrection will occur in a new construct.

Time for the Creation

In several verses of the Qur'an it is mentioned that Allah created the Earth and the Heavens in six days. This can be problematic as it seems to echo the Old Testament, which states that God created the Heavens and the Earth in six days and rested on the seventh day.

Firstly, the Qur'an refutes the idea that Allah became tired after the act of creation. One verse tells us that:

And indeed We created the heavens and the Earth and all between them in six Days and nothing of fatigue touched Us. [50:38]

{وَلَقَدْ خَلَقْنَا السَّمَاوَاتِ وَالْأَرْضَ وَمَا بَيْنَهُمَا فِي سِتَّةِ أَيَّامٍ وَمَا مَسَّنَا مِنْ لُغُوبٍ}
[ق:38]

Clearly, the statement "and nothing of fatigue touched Us" contradicts the statement in the Old Testament where God rests on the seventh day.

Secondly, one should be aware that the meaning of a 'day' in the Qur'an is much broader than the conventional meanings in use. This can be substantiated by the fact that the Qur'an refers to more than one type of day:

> *He sends the regulation of the affair from the heavens to the Earth, then on the day which is equal to one thousand years of yours, it will ascend to Him. [32:5]*

{يُدَبِّرُ الْأَمْرَ مِنَ السَّمَاءِ إِلَى الْأَرْضِ ثُمَّ يَعْرُجُ إِلَيْهِ فِي يَوْمٍ كَانَ مِقْدَارُهُ أَلْفَ سَنَةٍ مِمَّا تَعُدُّونَ} [السجدة:5]

In another verse we are told that:

> *The angels and the Rūh ascend to Him in a Day the measure whereof is fifty thousand years. [70:4]*

{تَعْرُجُ الْمَلَائِكَةُ وَالرُّوحُ إِلَيْهِ فِي يَوْمٍ كَانَ مِقْدَارُهُ خَمْسِينَ أَلْفَ سَنَةٍ} [المعارج:4]

Hence, there is one type of day that refers to the period between two successive sunsets; this is the day the Arabs were counting time with. There is also another type of day which is 1,000 years of what the Arabs were counting with (the lunar year which is 354.366 days long, not the tropical year of 365.2422 days), and another type which is 50,000 years long, with no specification of the length of one year.

Consequently, one cannot assume that the days of the creation mentioned in the Qur'an are necessarily the conventional unit of days as understood by humans. In this context, the Qur'an is pointing to the chronological order of the creation in the above verses, made especially clear in the verse [70:4]. Incidentally, past commentators of the Qur'an have interpreted the 50,000 years as being the length of doomsday.[29]

---

[29] See for example the commentaries of al-Tabari and Ibn Kathir.

Chapter Six

The Grand Design

From a modern scientific view we understand that the vast space containing the heavenly bodies is the volume that covers what we now call the universe. Using optical and radio telescopes, mankind is now able to explore the universe with a high degree of accuracy, covering distances which extend beyond billions of light years. Astronomers are able to analyse the light coming from luminous objects located at these distances and learn about their chemical composition in great detail. Currently, great efforts are being made to improve upon these techniques, such that information received has introduced new facts which confirm the existence of an interconnected creation and make for a balanced evaluation of every single part of the universe.

The question of mankind's position in the universe has gained increasing attention over the last few decades, especially following the discovery of the peculiar adjustments required by the laws of nature in order to make the existence of humankind on Earth explainable by the laws of physics. The possibility of developed alien life systems existing in another part of the universe seems to be remote. Despite the fact that we are now sure of the existence of other planets belonging to extrasolar systems within our galaxy and other galaxies, we are uncertain of whether they are capable of harbouring any life forms.

Scientific speculation offers high expectations for life, but the extremely difficult conditions required for developing complicated life systems like ours makes the probability of finding such a planet very low.[30]

Nevertheless, we are now sure of one fact: the construction and development of the whole universe is firmly connected with the possibility of developing life somewhere in this universe. A profound mediation both in experimental results and theoretical explanations has been set for the possibility of these discoveries, showing that all the parts are connected together, and suggests that there are some strange similarities and precise relationships between the smallest and the largest in this universe. This indicates a system that has been formed by a sort of cosmic conspiracy in order to make the existence of humans in such a universe possible. To gain an idea about what I mean here, I shall review some of the relationships between the basic constituents of the universe and the major forces at play.

The Neutrino

A neutrino is a neutral particle with no electric charge and is one of the basic constituents of the atomic nucleus. It was assumed to exist when studies conducted on the products of interactions of elementary particle showed that the equations could not be balanced without allowing for its presence. Its properties were discovered as a result of particle accelerator experiments whereby subatomic particles collided at high speeds, causing them to break up, such that their internal structure and contents could be investigated.

Elementary particles appear as traces on detection devices such as the cloud chamber, where steam molecules condense along the path of the

---

[30] Meadows, V. S., 'Planetary Environmental Signatures for Habitability and Life,' in Mason, J. (Editor), *Exoplanets: detection, Formation, Properties, Habitability*, Springer, (2008).

charged particle. Uncharged particles do not show such traces, making it difficult to discover them directly. This is why researchers were unable to prove the existence of the neutrino. However, they were able to show that these particles—which were formed immediately after the Big Bang—make up a large proportion of particles in the universe. In fact, it is estimated that there are about 109 (i.e. a thousand million) neutrinos for each proton and electron in the universe.

It is believed that the interaction of these particles with other materials is very weak. For example, the Earth is almost transparent to them, allowing them to pass through it easily. However, due to their vast numbers, the whole structure of the universe is very sensitive to their presence. It was generally believed that neutrinos were massless particles (i.e. energy waves), so physicists assumed that they moved as fast as light. Experiments conducted at the beginning of the 1980s showed that neutrinos might have some rest-mass, which is estimated at a few parts in 10-37 kilograms (i.e. 5 parts in 10 million parts of the mass of the electron). When we know that the electron is the smallest mass particle we can appreciate just how small the mass of a neutrino is. The high-density of these particles (about one billion particles in each cubic meter) means that their total mass is more than the mass of all the stars in the universe.

In order to understand how the mass of such a tiny particle affects the structure of the early universe, we can say that if the mass of these particles was a little greater, it could make a substantial change to the speed of universal expansion or completely halt the expansion, or even cause the universe to contract before the existence of man on Earth became possible. On the other hand, if the mass of the neutrino was even heavier, they would have gathered at the centre of galaxies because their escape velocity would be greater than their normal speed, forming a kind of heavy fog which would obstruct the rotation of the galaxy and halt its universal movement.[31]

---

31 Davies, P.C.W. *The Accidental Universe*, 61.

The calculations, made by a team of theoretical physicists at Texas Austin University in the United States, confirmed that any simple change in the neutrino mass would cause a serious cracking in the structure of the galaxy. In addition to all this, the neutrino plays a very important role in balancing the energies and spin of nuclear particles during their interactions, thus playing a highly sensitive role in defining the ratio of these particles in the structure of the universe from its early beginnings. If the power of these interactions was a little larger or a little smaller, then there would not have been enough hydrogen in the universe.

Hydrogen

Hydrogen plays a significant role in the chemistry of the universe; without it there would not be any organic materials or water on Earth or anywhere else in the universe, nor would there be enough fuel for burning stars such as the Sun. Consequently, there would be no life in the universe.

The normal hydrogen atom consists of one proton and one electron but there are two other hydrogen isotopes: deuterium with one proton, one electron and one neutron in its nucleus, and tritium with one proton, one electron and two neutrons in its nucleus. It is known that the deuteron (a deuterium nucleus) is an important component of the nuclear fusion cycle which occurs inside the Sun. This fusion reaction produces helium and an excess of energy which makes the Sun shine.

The hydrogen isotopes have the same chemical properties as normal hydrogen but they are different in their physical properties. The mass of the neutron is larger than the mass of the proton by about $1/1000$ of the proton mass. This is why the free neutron can decompose into a proton, an electron and anti-neutrino. If the difference between the mass of the neutron and the mass of the proton was any smaller (e.g. one third of its value), then the free neutrons would be unable to decompose into protons as they would have an insufficient mass to generate the necessary electron with which to balance the charge. This would lead to a fundamental change in the quality of the nuclear interactions in the universe. If the neutron mass was 99.8% of its current value, a free proton would decay into a neutron and a positron (positive electrons) and consequently, there would not be any hydrogen atoms in the universe. This would mean, again, that there could be no possibility for life as we know it.[32]

Nuclear Forces

Nuclear forces play a serious role in the structure of the nuclei of chemical elements and indeed, of the whole universe. The helium nucleus contains two protons and one neutron (helium-3) or two neutrons and two protons (helium-4), which is the dominant helium isotope in the universe. Since similar charges repel each other, the electrical force between two protons in an atom would push them away from each other with a tremendous force which amounts to more than ten million billion tons. However, nuclear forces bind the protons and neutrons found within the minute space inside the atomic nucleus because they work at very short distances. As such, nuclear forces are much stronger than the electrical forces; strong enough, in fact, to easily overcome the forces of the electrical repulsion. Because of their short range, nuclear forces act on neighbouring particles while electrical forces act on all charged particles due to their long range.[33]

This means that any proton in the nucleus is bound to its neighbour by nuclear forces which pull it to the nearest distance they can, while it is simultaneously being pushed away by the farthest protons with their electrical repulsion. In light nuclei which contain a small number of protons, the repulsion powers have a negligible effect; however, in the large nuclei that contain a large number of protons, electrical repulsion

---

[32] See for example, Paul Davies, *The Accidental Universe*.
[33] Povh, B.; Rith, K.; Scholz, C.; Zetsche, F. *Particles and Nuclei: An Introduction to the Physical Concepts.* Berlin:Springer-Verlag, (2002), p.73

is much more effective. This causes heavy nuclei to disintegrate spontaneously, producing radioactivity. This way the nucleus rids itself of the heavy stuff inside to become lighter and more stable.

If the nuclear forces were a little weaker, there would be only a small amount of stable chemical elements. If the value of any nuclear coupling constant is half of its current value, then the nucleus of say, iron, or even carbon, would not be stable for very long. This would lead to the loss of one of the most important elements for life, since carbon is an essential element for forming DNA (deoxyribonucleic acid), the molecule which contains the genetic code (as chromosomes). In short, the propagation of life would be impossible.

What would happen if the nuclear forces which hold the nucleus together were cancelled? The answer is that the universe as we know it would vanish, and instead, turn into flying protons and neutrons. In addition, there would be a very significant amount of hydrogen and no helium, i.e. the universe would be a silent, static and meaningless place, with no possibility of producing life.

Carbon and Oxygen

Since carbon and oxygen are essential elements in the cells of living organisms, the British astrophysicist Fred Hoyle gave great attention to understanding how these elements were formed in stars. Carbon is generated from combining three helium atoms, but this harmonious combination would be naturally rare without the necessary balanced arrangements. This is because the fusion of two helium nuclei produces an unstable beryllium nucleus. The possibility of a third helium nucleus fusing to make a carbon atom before the unstable beryllium nucleus decays depends on the energy by which the helium nucleus hits the temporary beryllium nucleus. This is due to the nuclear resonance requirement. When the quantum wave function of the helium nucleus becomes compatible with the internal vibration of the unstable beryllium nucleus, the cross-section to combine with the third helium nucleus will sharply rise.

Incidentally, the thermal energy of the nuclear constituents of a typical star is located exactly at the point of resonance in the carbon atom.[34] This wise adjustment of the natural properties results in a low probability resonance, such that sufficient carbon atoms are produced inside the stars. Without this, the abundance of carbon in the universe would be very low (around 1% of the present abundance) and no carbon-based life would be formed.

Carbon can be converted into oxygen during nuclear reactions, but because the nuclear resonance in an oxygen nucleus is well below the thermal level of the constituents, most of the carbon is saved from this fate. Nuclear reactions are complicated; however, the locations of nuclear resonances depend on the primary strength of fundamental forces in the universe, especially nuclear and electromagnetic forces. If the strength of these forces is different, the precise arrangements in the resonance of carbon and oxygen would not happen and the possibility of life on Earth would be negligible.[35]

Does this mean that life depends only on carbon? Absolutely not; scientists proposed other alternatives for life depending on which other elements were utilised, e.g. nitrogen—although, of course, this would result in a new type of creation. This means that creating life in another form or via a different developmental path would require a huge change in the universe, which no one is capable of except the One who performed the first creation.

---

34 Salaris, Maurizio; Cassisi, Santi, *Evolution of stars and stellar populations*, John Wiley and Sons, 2005, 119–121.

35 Ostlie, D.A. & Carroll, B.W. *An Introduction to Modern Stellar Astrophysics*, San Francisco: Addison Wesley, 2007.

The Divine Word and the Grand Design

Gravitational and Electrical Forces

It has been noted that gravitational force has universal domination due to its tremendously long-range effect—infinite, in fact. This force controls the universe and the movement of far galaxies to the extent that even light—which is so fast that it can circle the Earth seven times in one second—needs billions of years to reach us from there. The electromagnetic forces which dominate the microscopic world of atoms and molecules are much stronger than the force of gravity. To demonstrate the difference between gravitational and electrical forces in the nuclear realm, we must mention that the electrical force between the electron and the proton in a hydrogen atom is 1040 times stronger than the gravitational force between them!

The strength of these two forces depends on the fine structure constant of each, where the strength of certain forces reflects the value of this constant (commonly denoted α). The value of the constant with respect to gravitational force is $5.9 \times 10{-39}$, while its value with regards to electromagnetic power is $7.3 \times 10{-3}$. In fact, the determination of these values has a great influence on the organisation of the universe, as well as in the creation of typical stars and their stabilization as they produce light and heat.[36]

A good example of this is the Sun which, without such precisely determined values for the strength of the forces, would not be viable in its current form. The structure of a star depends on its ability to spill out the heat from its core onto its surface through radiation. In massive stars such as the so-called blue-giants, radiation energy becomes the dominant energy, with thermal energy dissipating from them through radiation.

If the mass of the star is smaller, this way of moving energy is not feasible because radiation cannot travel fast enough to maintain the surface of the star with sufficient heat. This is an important point; if there are

---

36 Davies, *The Accidental Universe*, p.48.

an inadequate number of ions on the surface of the star, it will become unstable and its heat removed by convection.

The convectional excitation of heat will be complemented by the flow of radiation energy and will prevent the temperature from being reduced substantially below the ionization temperature. This is why stars which use thermal convection methods are smaller and cooler than blue-giants. These stars are red-giants. The Sun—categorised as a yellow-dwarf—and some other stable stars fall between the bands of these two classifications.

The Accident and the Purpose

Brandon Carter, a famous astronomer, proved that if gravitational forces were a little weaker, electromagnetic forces a little stronger, or electron mass a little smaller in proportion to proton mass, then all stars would not be formed since insufficient temperature is available at the core and they would become red-dwarfs. Accordingly, a change in the opposite direction would turn all the stars into blue-giants, and as a result, the universe might not contain planets due to slightly lower gravity. In both cases, if gravity was a little more or a little less, then our universe would be totally different.

The facts introduced above have attracted the attention of scientists working in cosmology, atomic physics and astronomy, to the possibility that an intelligent power may stand behind the creation of the universe. Some scientists find it naïve to adopt the idea that its creation was purely accidental. The probability of an accidental universe being created is extremely low, and to say that many accidental chances have occurred coincidentally to make this universe possible is unreasonable. Among the numerous options, the possibility of the emergence of life—particularly intelligent life—is significantly low. How could all these constants and factors have been arranged with such values as to lead to the emergence of intelligent life and a level of consciousness that we enjoy? What sort of conspiracy is behind the scenes?

The Divine Word and the Grand Design

In the last two decades, the fact of the finely-tuned universe was exposed when theoretical research on cosmic microwave background radiation revealed the very accurate tuning of the initial conditions for creating this universe. The expansion rate of the universe depends on the value of the cosmological constant and this in turn depends on the balance between the respective contributions of the quantum vacuum and the bare cosmological constant. I have personally worked on these topics and am aware that the balance between the vacuum energy and the bare cosmological constant can be up to 10-50.

From the above, and many other observations, scientists realised that there was a strong relationship between every single part in the universe and the existence of humans on an Earth that moves around one specific star out of thousands of millions of stars in the whole universe. Yes, there might be other planets which harbour life, but none have been discovered as of yet.

The fact that we are here in this universe in spite of the odds stacked against us, induced scientists to attempt to find an explanation for the coincidences which make our existence possible. I admit that this is one of the hardest questions that one may face, and many geniuses in the scientific world have expressed their astonishment as to how our mental ability can be in such harmony with the universe that we are able to comprehend it perfectly. Einstein is quoted to have said: "The most incomprehensible thing about the universe is that it is comprehensible."[37]

The strong relationship between how the universe began and our highly developed and complex presence introduced a new concept into the mix, known as the anthropic principle. This notion emerged as an explanation for the observed constancy of nature and the harmony of the laws of nature with the emergence of intelligent life. Scientists have expressed two main views about this principle.

---

37 Einstein, A. *Physics and Reality, in Ideas and Opinions,* trans. Sonja Bargmann New York: Bonanza, 1954, p.292.

The Weak Anthropic Principle

According to one version of the anthropic principle, our existence is just a happy, rare chance by which sporadic natural incidences have combined over a very long period of time to make our existence possible. This is the weak anthropic principle, defined by Brandon Carter as follows: "What we can expect to observe must be restricted by the conditions necessary for our presence as observers."[38] Followers of this explanation admit that it was a rare opportunity to have intelligent life developing on Earth but they consider it some sort of factual reality that we are here. Put simply, this means: if such a 'happy chance' did not occur, we would not be here to talk about it. This can be gleaned from the words of Paul Davies when he said that Hoyle's example of the formation of carbon in stars through the right balance of resonance and energy "does not explain the coincidence of nuclear energies, but merely comments on the extreme fortune of the circumstances: had it not been so, we should not have been here to discuss it."[39] Yet here Davies is overlooking the fact that Hoyle's illustration involves bringing in the two requirements, which are not inter-dependent (that is, resonance and the star's temperature), in order to achieve the goal of producing enough carbon necessary for our existence.

Other scientists such as Steven Weinberg have tried to devalue the importance of Hoyle's remark by overestimating the probability of achieving the required resonances by claiming that it is not as rare as might be thought. Nevertheless, Weinberg finds no way of escaping the fact of fine-tuning on other occasions, like the precise adjustment of the value of the cosmological constant.[40]

---

38 Brandon Carter, *Confrontation of Cosmological Theories with Observation* (ed. M.S. Longair, Reidel, Dordrecht), 1974.

39 Davies, The Accidental Universe, 119.

40 Weinberg, Steven. 'The cosmological constant problems' in *Sources and Detection of Dark Matter and Dark Energy in the Universe*, Berlin: Heidelberg 2001, pp.18–26.

The Strong Anthropic Principle

Another version of the principle was proposed as evidence of deliberate organisation and design intended to make our existence possible. Brandon Carter called this the 'strong anthropic principle' and described it as thus: "The universe must be such as to admit the creation of observers within it at some stage."[41] This statement implies that the universe was tailor-made for habitation by creatures like us through the interplay of the laws of nature and the initial conditions, which arranged themselves in such a manner as to assure the presence of living organisms. Davies correctly recognises that: "The strong anthropic principle is akin to the traditional religious explanation of the world which says: that God made the world for mankind to inhabit."[42]

Many scientists showed support for the notion of the strong anthropic principle, among them Josef Silk, John Wheeler and John Barrow. Silk remarked that: "Gravitational instability and fragmentation must lead from giant clusters to galaxies to stars, and ultimately to planets and an environment suitable for the development of life."[43] Wheeler points to the fact that such a large universe is necessary for our existence and unless the radius of the cosmic horizon is more than 109 light years, the universe will collapse. John Barrow and Frank Tipler jointly authored a large volume in which they supported the strong anthropic principle, pointing to the coincidences of the physical values and the action of the laws, saying that: "The possibility of our own existence seems to hinge precariously upon these coincidences. These relationships, and many other peculiar aspects of the universe's make-up, appear to be necessary

---

41 Carter, B. "Large Number Coincidences and the Anthropic Principle in Cosmology", IAU Symposium 63: *Confrontation of Cosmological Theories with Observational Data,* Dordrecht: Reidel, 1974, pp291–298.

42 Davies, *The Accidental Universe,* 121.

43 Josef Silk, Cosmogony and the magnitude of the dimensionless gravitational coupling constant, *Nature,* 265, 710, 1977.

The Grand Design

to allow the evolution of carbon-based organisms like ourselves."[44] The primary remarkable fact is that our existence is conditional upon a delicate physical relationship and the values of certain fundamental physical constants, which occur through infrequent chance.

There are two possible explanations for why such a rare chance would take place; one is that the universe has been around for an infinite amount of time, which allowed for the emergence of a high level of cosmic organisation. This might be thought of as the result of a stupendously unusual statistical fluctuation from the far more probable condition of featureless disorder. This was the explanation given by Ludwig Boltzmann, who suggested that a cosmic-sized version of the fluctuations that produce a Brownian motion could create the optimum conditions required for those rare events necessary for our existence. This requires a universe whose age extends to a hundred billion or even billions of billions of years, which is out of the scope of present cosmological observations. The second option is to assume the existence of many universes (the multiverse proposal), out of which our universe happens to have enjoyed such rare conditions as to make our existence possible.

Davies has observed that: "Given the random cavorting of all atoms, wholesale cooperation of large numbers of atoms will conspire, after unbelievably long duration, to produce order spontaneously out of chaos."[45] A point to note here is that the explanation given by the weak anthropic principle is no explanation, since it does not offer any reason as to why such a happy chance would occur. Why should there be so much co-ordination to make our existence possible? On the other hand, the so-called strong anthropic principle is no real explanation either; it only states the fact that the delicate adjustments and fine-tuning of many parameters are purposely intended to have us present in this universe. But why? Why should there be observers at all? Wouldn't this

---

44 Barrow and Tipler, *Anthropic Cosmological Principle*, xi.
45 Davies, *The Accidental Universe*, 128.

indicate that the presence of an observer is necessary for some purpose? What then, is the purpose of this observer?

Explaining the Anthropic Principle

Fine-tuning is an observable fact; its connection with the existence of a conscious, intelligent observer who can comprehend it is the basis of the anthropic principle. However, there must be a reason for this fine-tuning; it indicates that some coordination took place to adjust the universal constants and the fundamental physical quantities in order to achieve the precise values which make our existence in this world possible. Why should there be such precise fine-tuning? Perhaps it would be sensible to say that this fine-tuning is a necessary pre-condition for managing the world with 'blind laws' rather than 'smart angels'. Should such fine-tuning not be available, the blind laws of nature[46] would not have been able to develop the universe sufficiently enough to have us here. Precise definitions, strict ruling and detailed instructions are always needed whenever you have a group of simpletons at work.

In this case, the universe has two options:

1. Have no freedom for the blind laws at work, thus making them work deterministically (which is in line with the thinking of Richard Dawkins).
2. Allow these laws some freedom to become indeterministic.

With the first option we end up with a world ruled by the deterministic laws of nature and by which the universe will need no driver/coordinator (i.e. no God). But this only allows for limited diversity of creation and is against what quantum mechanics tells us. The second option allows for the universe to be ruled by indeterministic laws that

---

46 By the 'laws of nature' I mean all the natural phenomena. These are different from the laws of physics which are normally our explanation for the laws of nature. See: Basil Altaie, *God Nature and the Cause*, KRM, 2016, Chapter 2.

work with some freedom, as John Polkinghorne expected; however, this option necessitates the presence of a driver/coordinator to choose from the different contingent possibilities and produces a widely diversified world. The presence of the intelligent observer would justify the need to verify and testify to the grand purpose.

This explanation shows that, despite the fact that we have blind innate laws at work in the world, the watchmaker, after all, is not blind. It is not the tools that are making the parts of the watch and collecting them, but it is the one using these blind tools who is the maker of the watch. These blind tools can see through their user, in a way that Dawkins couldn't recognise.

The Multiverse Hypothesis

The notion of a multiverse is a convenient concept that lends itself to explain anything that one may find peculiar. By its own construct, being dependent on infinite possibilities, this notion allows us to choose whatever we like out of an unlimited set. Steven Weinberg,[47] for example, uses it to explain the value of the cosmological constant, which he relates to anthropic issues. Similarly, Martin Rees[48] employs it to explain the whole set of anthropic coincidences, that is, to explain why our universe is a congenial home for life. Others such as Barrow[49] use it to speculate about the values of physical constants in our universe. In short, the notion of a multiverse, as presented in the arena of science today, makes everything possible.

The basic idea of having multiple universes seems completely natural; we have many planets, many stars, many galaxies and many galactic clusters, so why have only one universe? There are three origins for the

---

47 Weinberg, S., *The Cosmological Constant Problems*, see footnote (28).

48 Rees, M. J. 2001, *Just Six Numbers: The Deep Forces that Shape the Universe* (Basic Books). Rees, M.J. 2001 *Our Cosmic Habitat*, (Princeton University Press). M J Rees 2001, 'Concluding Perspective', astro-ph/0101268.

49 Barrow, J. D. and J.K. Web, Inconstant Constants, *Scientific American*, June 2005.

notion of a multiverse: the first is the Everett-Wheeler interpretation of quantum mechanics, the second is chaotic inflation proposed by Andre Linde[50] and the third is string theory.

The general theory of relativity uncovered the integral nature of space and time as being interwoven into one spacetime continuum. Accordingly, we should have one universe. However, inflation theory suggests that there could be multiple, causally disconnected regions of spacetime, each forming a manifold—therefore, a universe on its own. This is the origin of the multiverse notion, derived from the theory of inflation.

Everett proposed that quantum states split or branch upon measurement as a means to resolve the problem of measurement in quantum mechanics. This proposal triggered the notion that each branch is a universe on its own, thus producing a multiverse (many worlds) after a sequence of measurements.

String theory brings in a similar idea, suggesting that the temporal development of strings in higher dimensions would produce superstrings (or branes) which express the character of a multiverse.[51]

It is important to define the term 'multiverse'. This question has been raised by George Ellis, the renowned cosmologist and leading expert on Einstein's general relativity, who has written several articles on evaluating the notion of a multiverse. Ellis says: "There is however a vagueness about the proposed nature of multiverses."[52] In a recent article published in *Scientific American*, he analysed the proposal of a multiverse and concluded that: "We are going to have to live with that uncertainty. Noth-

---

50 Linde, A. D. 1983, Physics Letters 129B, 177; Linde, A. D. 1990, *Particle Physics and Inflationary Cosmology*, Harwood Academic Publishers, Chur, Switzerland.

51 Richard J Szabo, *An Introduction to String Theory and D-brane Dynamics,* Imperial College Press; 2 edition, 2011.

52 George F.R. Ellis, Ulrich Kirchner, William R. Stoeger, *Multiverses and Physical Cosmology,* 347, 2004 921–936.

ing is wrong with scientifically based philosophical speculation, which is what multiverse proposals are. But we should name it for what it is."[53]

Astronomers have realised that what they see with their most powerful telescopes is not the whole universe, but only part of it. This can extend, at most, to 42 billion light years. This is our cosmic visual horizon. But there is no reason to believe that the universe stops at such boundaries. Beyond it could be many more domains similar to what we can see. Each might have a different initial distribution of matter, but the same laws of physics would operate in all. Most cosmologists today accept this type of multiverse, which Max Tegmark calls 'Level 1'.[54] Yet some physicists like Alexander Vilenkin go much further in painting a dramatic picture of an infinite set of universes with an infinite number of galaxies, an infinite number of planets and an infinite number of people with your name who are reading this book. Needless to say that these infinite universes are but mere possible mathematical solutions endorsed by equations which set the formulation for the proposed systems. Since the dawn of mathematics and algebra we have known that in many cases, an algebraic equation can have an infinite number of solutions. Physicists usually pick the solutions which represent realistic states and ignore the others. The fact that general relativity has enabled us to write an equation for the whole of spacetime has given some physicists the excuse to interpret different solutions as representing different universes that may exist. But the bizarre thing is that they are providing these solutions without applying any constraints, since they have none to apply. Accordingly, they suggest completely different types of universes with different physics, different histories and maybe even different numbers of spatial dimensions. Most will be sterile, although some will be teeming with life. This is called a 'Level 2' multiverse.

Similar claims have been made since antiquity by many cultures. What is new is the assertion that the multiverse is a scientific theory,

---

53 George, F.R. Ellis, "Does the Multiverse Really Exist?" *Scientific American*, August 2011, 39–43.
54 Max Tegmark, *The Multiverse Hierarchy*, 2009.

with all its implications being mathematically rigorous and experimentally testable. Even so, George Ellis says: "I am skeptical about this claim. I do not believe the existence of those other universes has been proved—or ever could be."[55]

There are various proposals as to how such a proliferation of universes might arise and where they would all reside. They might be sitting in regions of space far beyond our own, as envisaged by the chaotic inflation model of Alan Guth, Andre Linde and others.[56] On the other hand, Paul Steinhardt and Neil Turok suggest that such multiverses may exist at different epochs within the cyclic universe model of universe.[57] David Deutsch advocates that they might exist in the same space as us but in a different branch of the quantum wave function.[58] Conversely, Max Tegmark and Dennis Sciama believe that they might not have a location, being completely disconnected from our spacetime.[59]

For a cosmologist, the basic problem with all multiverse proposals is the presence of a cosmic visual horizon. The horizon is the limit of how far away we can see, because signals that have been travelling at the speed of light—which is finite—since the creation of the universe have not had enough time to reach us. All the parallel universes lie outside our horizon and remain beyond our capacity to see, no matter how much our technology evolves. In fact, they are too far away to have had any influence on our universe whatsoever. This is why none of the claims made by multiverse enthusiasts can be directly substantiated.

---

55 George Ellis, "Opposing the Multiverse", *Astronomy & Geophysics*, Volume 49, Issue 2, 1 April 2008, 2.33–2.35,

56 Andrei Linde, "The Self-Reproducing Inflationary Universe", *Scientific American,* November 1994.

57 Gabriele Veneziano, "The Myth of the Beginning of Time", *Scientific American*, May 2004.

58 David Deutsch and Michael Lockwood, "The Quantum Physics of Time Travel", *Scientific American*, March 1994.

59 Max Tegmark and Denis Sciama, Parallel Universes, *Scientific American*, May 2003.

Let us look at the landscape of the planets in our universe. Suppose we attempt to extrapolate from their properties and the laws that apply to their control and development. Do we see that each planet has its own laws? No. If we examine the landscapes of the billions of galaxies in our universe, do we find different laws for each of their formations and developments? No. Why then, should we allow for different laws of physics to be at work within other worlds and landscapes within the multiverse? It seems to me that such an adventure is unnecessary.

Looking through the recent defences for the multiverse theory, one can see that some proponents of the notion have become more modest regarding their speculations. Max Tegmark says that no-one from the proponents of the multiverse claims that these parallel universes really exist.[60] Therefore, let us then rely on objectivity and stick to the fact that these are mere mathematical solutions to resolve the question of whether they exist or not. Objectivity tells us that we should not scrutinise speculations too closely once we have set a basis which allows us to take everything from our imaginations as possible. The open landscape of the multiverse is dangerous as it may take science into a new venture that goes beyond the logic by which we comprehend reality. We have to admit that objectivity tells us that the multiverse of Level 1 can only be accepted as a possibility which we can comprehend. In no way can we deal with other levels since our knowledge is not equipped with the necessary ingredients required for such a venture.

If we are to tackle such questions, we need to break away from what we know and try to investigate such possibilities systematically. For example, we may think of a world with antigravity or a world working with an inverse of the second law of thermodynamics. We can play with variances of the laws of statistics and quantum mechanics in order to see

---

60 Max Tegmark, Are Parallel Universes Unscientific Nonsense? Insider Tips for criticizing the multiverse, *Scientific American,* February 2014.

what kind of outcomes we obtain. In this case, the multiverse hypothesis could be a productive research programme.

From an Islamic point of view, there is no contradiction between belief in Islam and the existence of a multiverse of any type, since it is assumed that Allah the Creator is able to create whatever He wills. Therefore, there is no limit on the number of worlds that He can create and no restrictions on their properties. In fact such worlds, even with different laws of physics, are deemed necessary to acknowledge the afterlife, be it heaven or hell. In some literature and religious narrations there is a detailed description of such worlds. Thus, in one way or another, the notion of a multiverse may help give a scientific touch to some metaphysical religious thoughts.

In sum, although the idea of parallel universes is attractive, it is still immature and unfit to answer many questions. Hereafter, I would like to present the Islamic understanding for the fine-tuning of the universe.

Fine-Tuning and *Taskhīr* in Islam

In Islam, we encounter a principle which states that whatever we see in the universe was arranged for our service and to make the purpose of our presence lucid. This is called the principle of subservience (taskhīr) and means that the design of the universe has been purposefully created for the service of humans and to make our lives possible.

The Qur'an gives accurate descriptions for the construction of the world and the introduction of an intelligent creature who can contemplate and comprehend the universe in order to learn about the laws governing it. We first read that the world was created with the utmost care and that the values of its constituents were chosen with great accuracy:

And everything with Him (Allah) has a measure. [13:8]

{وَكُلُّ شَيْءٍ عِنْدَهُ بِمِقْدَارٍ} [الرعد:8]

He has created everything, and has measured it exactly according to its due measurements. [25:2]

{وَخَلَقَ كُلَّ شَيْءٍ فَقَدَّرَهُ تَقْدِيرًا} [الفرقان:2]

Surely, We have created everything according to a measure. [54:49]

{إِنَّا كُلَّ شَيْءٍ خَلَقْنَاهُ بِقَدَرٍ} [القمر:49]

These verses and many others assert that the creation was performed to a measure. This allows us to assume that such a measure is necessary for achieving a specific target. Furthermore, it implies that the creation was achieved according to certain rules that would require delicate measures and care.

The Qur'an reveals that the Creator made everything on Earth and in the Heavens subservient to mankind:

And He has made subservient (subjected) to you whatsoever is in the heavens and whatsoever is in the Earth, all, from Himself. Surely there are signs in this for a people who reflect. [45:13]

{وَسَخَّرَ لَكُم مَّا فِي السَّمَاوَاتِ وَمَا فِي الْأَرْضِ جَمِيعًا مِّنْهُ إِنَّ فِي ذَٰلِكَ لَآيَاتٍ لِّقَوْمٍ يَتَفَكَّرُونَ} [الجاثية:13]

See you not that Allah has made subservient (subjected) to you whatever is in the heavens and whatever is in the Earth, and granted to you His favours complete explicitly and implicitly? And among men is he who disputes concerning Allah without knowledge or guidance or enlightening book. [31:20]

{أَلَمْ تَرَوْا أَنَّ اللَّهَ سَخَّرَ لَكُم مَّا فِي السَّمَاوَاتِ وَمَا فِي الْأَرْضِ وَأَسْبَغَ عَلَيْكُمْ نِعَمَهُ ظَاهِرَةً وَبَاطِنَةً وَمِنَ النَّاسِ مَن يُجَادِلُ فِي اللَّهِ بِغَيْرِ عِلْمٍ وَلَا هُدًى وَلَا كِتَابٍ مُّنِيرٍ} [لقمان:20]

And He has made subservient for you the night and the day and the Sun and the Moon. And the stars are made subservient by His

The Divine Word and the Grand Design

> command. Surely there are signs in this for a people who understand. [16:12]

{وَسَخَّرَ لَكُمُ اللَّيْلَ وَالنَّهَارَ وَالشَّمْسَ وَالْقَمَرَ وَالنُّجُومُ مُسَخَّرَاتٌ بِأَمْرِهِ إِنَّ فِي ذَلِكَ لَآيَاتٍ لِقَوْمٍ يَعْقِلُونَ} [النحل:12]

Indeed, these are signs for people who understand. It is worthy to note here that whenever the Qur'an mentions subjecting or subservience, it mostly comes in the context of the general creation of the Heavens and the Earth. So, the intended subjection of the heavens, stars, the Sun and the Moon is related to the issue of creation, in which the organisation of creation is to be at man's service. To say that the appearance of mankind at a particular age in the universe is just a 'lucky moment' is a superficial understanding of the anthropic principle.

In fact, there are many other verses in the Qur'an which mention subjecting the rivers and the winds and water falling down to the ground, all for the benefit of mankind. This is part of the subservience (taskhīr) extended for the life of humans. But why should humans be so important?

The answer lies in the fact that the human is destined to be a very different creature from all other creatures on Earth; he enjoys a free will, the ability to choose and the ability to perform willful acts. All of this is to test this creature and see how he will use his resources and how he behaves:

> Who created death and life that He might try you—which of you is best in deeds? And He is the Mighty, the Forgiving. [67:2]

{الَّذِي خَلَقَ الْمَوْتَ وَالْحَيَاةَ لِيَبْلُوَكُمْ أَيُّكُمْ أَحْسَنُ عَمَلًا وَهُوَ الْعَزِيزُ الْغَفُورُ} [الملك:2]

Such a test requires that there should be an extension of life through an afterlife, with either reward or punishment, and this is exactly what the Qur'an purports. However, these explanations may be metaphors for reward and punishment, since the descriptions given in the Qur'an cannot be comprehended within our physical and mental scope. Therefore we cannot take the descriptions given literally. Furthermore, the terms 'reward' and 'punishment' are far from being comprehensible, since they are part of a metaphysical state that humans are promised in the afterlife. Consequently, according to modern understandings in kalam, we should accept these theological narratives as they are without too much inconclusive speculation. In short, the duty of man is well-defined in Islam—to do good deeds and not be an infidel by engaging in mischief or polytheism:

> *Say, I am but a man like yourselves; but it is revealed to me that your God is One God. So let him who hopes to meet his Lord, do good deeds, and let him join no one in the worship of his Lord.*
> *[18:110]*

{قُلْ إِنَّمَا أَنَا بَشَرٌ مِثْلُكُمْ يُوحَى إِلَيَّ أَنَّمَا إِلَهُكُمْ إِلَهٌ وَاحِدٌ فَمَنْ كَانَ يَرْجُو لِقَاءَ رَبِّهِ فَلْيَعْمَلْ عَمَلًا صَالِحًا وَلَا يُشْرِكْ بِعِبَادَةِ رَبِّهِ أَحَدًا} [الكهف:110]

The scientific explanation for the rare chances that allow for our existence is fundamentally different from the theological explanation given by the Qur'an. Although the Qur'an calls upon us to contemplate creation and to search for its origins, we find that it is primarily concerned with proclaiming the greatness of the Creator—His uniqueness, oneness and dominance over all of existence. The presentation of the view given in the Qur'an is a very effective approach in exposing how the methodology of natural sciences differs from the methodology of religions. However, we should admit that the methods and goals of science ignore considering questions which are no less important than questions about our existence; questions which are related to our destiny. But we should also admit that a purposeless universe, acting in accordance with certain rules or laws of unknown origin and of unknown drive to produce all this diversity and complex organisation, is something that raises more questions than it solves. Perhaps it is because our mental capabilities are not yet equipped to grasp such paradoxes or that the messages from the

Divine were sent to us in order to provide a guide to help us seek our quest in a correct and more efficient way.

Purpose for the Existence of Humans?

Believers in the strong anthropic principle are still unsure of the supreme goal and main purpose for the existence of humans despite knowing the truth behind the anthropic principle and its purpose. The absence of a great purpose for the existence of humans on Earth would make the anthropic principle no more than a mere coincidence. Humans, from a materialistic point of view, are part of the material universe—no more, no less. So why should the universe have been created to serve them? On the other hand, the validity of creating this universe for the service of humans would make the approach of some scientists, such as John Wheeler, acceptable—if somewhat bizarre. Wheeler states that: "If this is just the human being then what would the universe look like?"[61]

This approach leads to one conclusion: that there is an inevitable purpose for the existence of man. Others such as John Barrow and Frank Tipler adopted the same approach, maintaining that: "Our presence imposes serious limitations on the proportion of photons to protons in the universe."[62] This statement by Barrow and Tipler is vague and can only mean that there is an inevitable purpose for the existence of the human beings. Now, my question comes in this context: if the human being, according to the materialistic perspective, is nothing more than an insignificant part of this enormous universe, and if this mass of meat and bones that lives on Earth for a few decades will only end up as a pile of ash, then what is the ultimate goal of man's existence?

The supreme consciousness and the advanced mental capabilities of humans cannot be just wasteful development, be it a result of natural Darwinian evolution or careful design and creation. Therefore, the purpose

---

61 Panda, N.C. *Vibrating Universe*, Motilal Banarsidass; 1st edition, 1999.
62 Barrow and Tipler, *Anthropic Cosmological Principle*.

of having humans on Earth cannot be comprehended through a solely materialistic and physical approach. We need to have some sort of metaphysical assumptions or our existence cannot be explained.

Human consciousness is a highly-valued channel that may work as a link between two worlds: the physical world and the metaphysical world that lies beyond our direct senses. It is thus important to recognise the value of human consciousness and to distinguish it from the inherent consciousness found in other creatures.

The Islamic understanding of the anthropic principle poses an integrated sense of addressing the purpose of the creation. It starts with Allah's Will to create and ends with a world that accommodates a creature who appreciates its metaphoric value and, consequently, appreciates the value of creation. Although we do not know the motives of the Creator, we may be able to understand them through investigating creation. Only then may we be able to probe the 'mind' of God.

Rational Muslim scholars (the mutakallimūn) tried to address the question of God's motive for creating the world and came to the conclusion that He could not be driven by a motive because, they argued, that if such a need existed, God would be lacking the self-sufficiency which is necessary for His omnipotence.

Humans have free will (at least in the apparent state of affairs) and limited abilities that they use within their physical and moral activities in order to achieve the aims of their existence on Earth. These aims and goals include: exploring the world, discovering how it developed, its movement and how it works. This is something humans have done ever since they became aware of themselves and the world. But what is the point of exploring the world, and what is the value and usefulness of the knowledge that is obtained?

The first goal is clear; humans invest this knowledge into improving their living conditions and cultivating their surroundings in order to overcome the difficulties of life. For example, when humans discovered fire they used it to cook their food. But what is the purpose of cultivation and welfare? There must be a higher purpose, and for this we need to view the world from a comprehensive and broader perspective. Was the goal of creating this world, full of atoms and galaxies and humans—with their extraordinary consciousness and marvellous mental abilities—only for the cultivation and wellbeing of all life forms? What is the point of this goal if an individual lives only for a few decades before dying and their existence is then completely obliterated? Certainly, our minds provoke us to think of a higher and more sophisticated purpose; a purpose with a deeper meaning and a broader impact than what I have already mentioned. To achieve the sublimity of this purpose and to be consistent with the sublimity of the world's creation, this purpose must be connected to the sustainability of existence, which means the sustainability of human consciousness. This is the result of man's existence in this physical world. If human beings are able to understand creation and discover its secrets, then they should be worthy enough to comprehend that they were created by a greater power who also created a world that is subject to them.

This fact will not be grasped directly by humans through solving differential equations, analysing the DNA from living cells or doing experiments, but it will be indirectly grasped when they greatly enhance their consciousness and add to it another dimension—the metaphysical one.

I believe that this transcending status can be obtained through intuition and not necessarily through rational reasoning, since the unseen is imperceptible directly through the mind, which is derived through extrapolations using intuitive knowledge. If one does this *and* invests his partial will and abilities well, then he will know Allah the Creator, the Omnipotent, and realise that his presence on Earth is not in vain and that the creation of the universe is not, and cannot, be in vain. Therefore, when man gets to know his Creator, he may achieve the ultimate goal of existence—to achieve the indefinite sustainability of their existence.

Thus, we can understand the anthropic principle in Islam in accordance with its integrated context, starting with Divine will. This is fulfilled at all stages of the universe's creation to provide conditions for man's existence and implement God's will according to His Divine norm, in which creation is shaped as one entity in all its multiple shapes. All of this is intended for human beings to know their Lord.

Thus, the anthropic principle as understood in Islam is part of a comprehensive and integrated system concerned with the issue of the beginning, the means and the purpose of existence. The basis of this purpose is to know Allah and to reach Him. For this to happen there must be a path for man to follow.

> O believers, keep your duty to Allah, and seek means of nearness to Him, and strive hard in His way that you may be successful. [5:35]

{يَاأَيُّهَا الَّذِينَ آمَنُوا اتَّقُوا اللَّهَ وَابْتَغُوا إِلَيْهِ الْوَسِيلَةَ وَجَاهِدُوا فِي سَبِيلِهِ لَعَلَّكُمْ تُفْلِحُونَ} [المائدة:35]

This may explain the reason behind man possessing limited will and ability, represented by his mental and physical capabilities, in order to achieve the purpose of his creation and development to such a stage. It is to get to know the Creator in a process of cosmic development and selection that will continually ascend until we become part of Him.

The relationship of the universal constants, both amongst themselves and with their relations in building the cosmic structure and its parts (large and small) according to the rules and principles that tolerate the presence of other alternatives as well, is not an isolated case. For example, carbon formation is essential for the existence of life but it is not the only requirement. It is possible to have life forms without carbon, as we mentioned earlier, and scientists have developed pathways for life forms that do not depend on carbon for their emergence and development. These forms of life will inevitably and certainly be different from ours, although the aim of creation would remain unchanged and the intermediary of creation (the successor) also remains. In addition, the anthropic principle also remains as an intermediary and a guide towards Allah. Hence, we can understand the meaning of the verse:

The Divine Word and the Grand Design

> *If He wills, He could dispose with you and bring forth a new creation. [35:16]*

{إِنْ يَشَأْ يُذْهِبْكُمْ وَيَأْتِ بِخَلْقٍ جَدِيدٍ} [فاطر:16]

Therefore, there is no confusion amongst Muslims regarding the anthropic principle. They are not faced with the question of whether the universe has been created for the sake of humans or whether *they* exist for the sake of the universe and its knowledge. Since Muslims know that humans are rational beings who were created to serve a purpose, it would be ludicrous to think that these rational beings could be servants to an irrational object. The Holy Qur'an has shed light in many places on this matter:

> *They worship beside Allah that which neither hurteth them nor profiteth them, and they say: These are our intercessors with Allah. Say: Would ye inform Allah of (something) that He knoweth not in the heavens or in the Earth? Praised be He and High Exalted above all that ye associate (with Him)! [10:18]*

{قُلْ أَتُنَبِّئُونَ اللَّهَ بِمَا لَا يَعْلَمُ فِي السَّمَاوَاتِ وَلَا فِي الْأَرْضِ سُبْحَانَهُ وَتَعَالَى عَمَّا يُشْرِكُونَ} [يونس:18]

Regarding parallel worlds, there is nothing in the Qur'an that argues against or completely negates the idea of the existence of other worlds. There is no theological restriction on the number of worlds that can be created by Allah, as long as this number is finite. The same applies to the finite number of atoms of the world. On defining finiteness, Muslim scholars relied on the following verse to verify their arguments:

> *He (Allah) keeps count of all things (i.e. He knows the exact number of everything). [72:28]*

{وَأَحْصَى كُلَّ شَيْءٍ عَدَدًا} [الجن:28]

To conclude, I should emphasise that Allah rules the world through well-defined blind laws which are formatted to be indeterministic, in order to allow for His intervention by choosing which of the possible events are to be affected. Blind laws are necessary to establish causal relationships, which are in turn necessary to establish law and order in the world, a basic requirement to recognise the Creator and Sustainer of the world.

Chapter Seven

Evolution

The evolution of living organisms is one of the most important issues in modern science. Its significance is primarily related to the understanding of the constitutive realities of living beings, their interactions, modalities, and the inter-relationships of these vital formations. This is crucial if we are to comprehend the mechanics of their development and the reason for the diversity we see in the world around us. This comprehension also enables us to learn how to deal with any emergent changes that may occur in their morphology or physiology and facilitates an understanding of any deficiencies present which might manifest as diseases and deformities.

Texts from the monotheistic scriptures provide us with a particular vision for the existence of mankind on Earth. We are told that the shape of the first man was moulded by God out of mud and blessed with His Spirit, before being transformed into a human being.

There have been a wide range of explanations for these religious texts; one of which is the literal embodied perception commonly implemented by the Jews in their interpretation of the sacred texts. They maintain that God made the first man out of wet mud, fashioned him and then breathed life into his body. However, some misunderstanding emerges

The Divine Word and the Grand Design

from their interpretation of the words 'God created man in his own image' from which they deduce that man was created in God's image. Islamic exegesis is replete with myths about the emergence of the universe and the source of life, which were mainly inspired by the ancients and their tales of how creation began. The majority of Islamic written heritage bears a similarity to the creation myths of the Old Testament as well as legends such as the separation of sky from water and the emergence of the Earth, as described in Genesis.

In Islamic literature we can spot some of these tales (some of which are attributed to the Prophet Muhammad ﷺ, while others are narrations of his companions such as Abu Hurayrah, Ibn ʿAbbas, and Kaʿb Al-Ahbar), in addition to the esoteric explanations that have burnished the story of creation as symbolic parables, rather than subject to changeable interpretations depending on variant signs and other symbols.

The quest to discover the origins of life and its manifestation on Earth is not only a challenge for the human mind, but also an adventure that might lead us to confront our traditional understanding of some of the principles of faith that we have inherited from our cultural heritage. This question has become one of the major contemporary philosophical issues of our time. In this chapter, I will present some of the challenges and intellectual problems that are related to the issue of the creation of man and his subsequent evolution over the ages.

Mankind has been unable to explore the depths of this issue in a serious manner until the recent development of theoretical, functional and exploratory tools, enabling us to shed light on the matter of the beginning of creation. This was developed by the English zoologist Charles Darwin and his contemporaries, although the idea was conceived by their predecessors. In the first half of the 19th century, Darwin recorded a large number of observations and documented the living organisms that he collected along a voyage covering many parts of the Earth. After studying these observations and documents, Darwin came to the conclusion that organisms have evolved biologically through the ages. The shapes and function of their organs, in addition to their capabilities, change according to the requirements of their surrounding environmental conditions. This is known as adaptation. If they cannot adapt, these creatures are not able to live and become extinct. Take the example of two types of butterflies: one can change its colours to be compatible with the environment in which it lives while the other cannot. Adapting provides the former butterfly with the ability to hide and camouflage itself from predators that might otherwise pounce upon it and eat it. Meanwhile, the other type of butterfly remains clearly visible and is easily attacked by its predators, generation after generation. Obviously, the second type will become extinct, whereas the first will prevail due to its ability to hide from predators.

According to Darwin's original theory, the evolution of living beings is a sequence of random events that occur to show new traits in subsequent generations. These traits, facing the environment, will be tested by success or failure. If a trait fails to stand the impact of the environment, the creature will perish. If it withstands the requirements of the environment, the evolving creature will survive. This interaction with the environment is called 'natural selection'.

Whenever an organism encounters adverse conditions, the evolution of its physical body and its physiological functions through subsequent generations will ensure its survival. If there is a positive development which eventually helps the organism to counter what was once invincible, then this kind will survive; however, if the new development lacks the required characteristics, then this species will vanish. The term 'survival of the fittest' arose from this principle.

Following this, Darwin decided to include all the patterns of life in a full scope of evolutionary biology. He claimed that all organisms have one origin from which other living species have been evolving as a series of ramified patterns, arising from one another as determined by genetic mutations. The structure of genetic heredity, discovered in the 1950s by Watson and Crick, allowed for a leap in the understanding of the dynamical structure of evolving, living organisms and formed a base for the accurate identification of the link between organisms. This discovery in particular supported Darwin's basic theory about the origin of species and paved the way for neo-Darwinism.

Darwin's theory of biological evolution, cast into the modern understanding, is based mainly on the assumption that there are two mechanisms at work: the first is the assumption of random genetic mutations taking place in chromosomes and the second is natural selection, by which nature chooses the 'fittest' of the offspring for survival, to the detriment of the weak. In other words, if the new qualities are unable to cope with the surrounding environmental realities, this will reflect the weakness of the living organism and result in extinction. Hence, the fittest organisms are solely adapted to survival on Earth, which highlights the notion of natural selection. As long as the struggle for survival is an ongoing process, the organic evolution of living organisms is a persistent methodology.

The question of organic evolution is an emerging issue that should be addressed within the framework of the new kalam. Therefore, this chapter will include an analysis of the question of biological evolution from both a theological point of view and as a scientific analytical argument.

Qur'an and the Evolution of Living Organisms

The first question that arises is whether or not biological evolution is in line with Qur'anic thought. It is commonly believed that the idea of organic evolution of man and living creatures is incompatible with religion. Relying on a literal understanding of religious scriptures, people have long supposed that living organisms were originally and directly created by God out of soil, and that man was created from clay and subsequently moulded into the form of a human being. Until recently, people believed that insects, worms and ants self-arose from non-living materials—a point of view adopted by Aristotle and many of the Elders of Greece.

Several verses of the Qur'an imply that God created man from clay (or dust):

> *(Remember) when your Lord said to the angels: "Truly, I am going to create man from clay." [38:71]*

{إِذْ قَالَ رَبُّكَ لِلْمَلَائِكَةِ إِنِّي خَالِقٌ بَشَرًا مِنْ طِينٍ} [ص:71]

> *Verily, the likeness of 'Iesa (Jesus) before Allah is the likeness of Adam. He created him from dust, then (He) said to him: "Be!"— and he was. [3:59]*

{إِنَّ مَثَلَ عِيسَى عِنْدَ اللَّهِ كَمَثَلِ آدَمَ خَلَقَهُ مِنْ تُرَابٍ ثُمَّ قَالَ لَهُ كُنْ فَيَكُونُ} [آل عمران:59]

> *O mankind! if ye are in doubt concerning the Resurrection, then you should know that We have created you from soil, then from nutfa (sperm), then from a clot, then from a little lump of developed and undeveloped flesh, that We may make (it) clear for you. [22:5]*

{يَاأَيُّهَا النَّاسُ إِنْ كُنْتُمْ فِي رَيْبٍ مِنَ الْبَعْثِ فَإِنَّا خَلَقْنَاكُمْ مِنْ تُرَابٍ ثُمَّ مِنْ نُطْفَةٍ ثُمَّ مِنْ عَلَقَةٍ ثُمَّ مِنْ مُضْغَةٍ مُخَلَّقَةٍ وَغَيْرِ مُخَلَّقَةٍ لِنُبَيِّنَ لَكُمْ} [الحج:5]

> *And of His signs is your creation from soil, and behold you human beings, spreading! [30:20]*

{وَمِنْ آيَاتِهِ أَنْ خَلَقَكُمْ مِنْ تُرَابٍ ثُمَّ إِذَا أَنْتُمْ بَشَرٌ تَنْتَشِرُونَ} [الروم:20]

> *Allah created you from dust, then from a little sperm, then He made you pairs (the male and female). [35:11]*

{وَاللَّهُ خَلَقَكُمْ مِنْ تُرَابٍ ثُمَّ مِنْ نُطْفَةٍ ثُمَّ جَعَلَكُمْ أَزْوَاجًا} [فاطر:11]

The concept of the original creation in these verses has a precise reference to Adam himself. Unlike Adam, all successive human beings have been created from sperm. These indications generated a necessity to understand creation as implied in the sacred scripts. It was taken to feature on two levels: the first referred to the first man (specifically Adam) whilst the second was intended to mean mankind in general. Here the process takes place through the food cycle, which originally consists of soil and water and turns the seed into a plant, nourishing people and animals. Thus, it was food that comprised of a sperm and an egg from which humans developed.

After a profound reflection of Qur'anic verses, one can observe the ingenuity of the Holy Qur'an with reference to the perception of creation. That is to say, it contains hidden meanings that can be subjected to more than one interpretation. In fact, the seventh verse of Surah Al-Imran explicitly states that the Qur'an contains verses that are firm and clear, whilst other verses are problematic or difficult to interpret (mutashābihat). The verses have a straightforward meaning when they handle issues related to sacred orders, religious orders and the laws of shari'ah. On the other hand, the verses which are problematic are those covering topics concerning cosmological arguments, such as the creation of man and the universe or describing metaphysical matters of creed. In this way, the Qur'an itself paves the way for exegetes to present their ideas and interpretations. However, some people extract meanings which may deviate from the intended meaning by polishing them with contrary indications. Accurate interpretations of Qur'anic verses always abide within the context and grammatical rules of the Arabic language. The Qur'an asserts this by stating:

> *We have revealed it—an Arabic Qur'an—that you may understand. [12:2]*

{إِنَّا أَنزَلْنَاهُ قُرْآنًا عَرَبِيًّا لَعَلَّكُمْ تَعْقِلُونَ} [يوسف:2]

Evolution

There are problematic verses in the Qur'an because it expresses the word of God in classical Arabic. Many of the sentences are structured in such a way as to hold a variety of clues, many of which are not interpretable by man, who has a limited cognition compared to the Almighty and His message. The more acquaintance one has with the Qur'an, the more meanings are exposed. The Holy Qur'an has been, and always will be, a perpetual truth.

It makes sense to say that it would have been inappropriate to explicitly disclose the details concerning creation and other sophisticated issues to a people who would have great difficulty in comprehending them and accepting them during specific eras of time. Had the Qur'an revealed these details, it would have required a lot of explanation and induction to cover many other fields of science which might not necessarily ensure a proper comprehension of the matters involved. Therefore the verses concerned with these subjects have maintained an ambiguity in which the general meaning dominates, but is bare of further description. The creation of the seven heavens and their destiny, for example, is open to much potential interpretation and the reader can deduce the intended implications within the context of the general meaning.

Another motive that impedes many people from accepting the idea of the organic evolution of living organisms and humans is their fear of diminishing the sole role of the Creator if they recognise the mechanism of natural evolution. They are wary of being accused of atheism. On the contrary, the unequivocal truth prescribes the following purview: that contemplation, not only of the correlation that knits the universe with the Creator but also of the natural mechanisms running the world (which seem to be axiomatic and spontaneous), confirms the unquestionable truth that Allah is the Living, the Self-Subsisting and the Eternal, and has power over everything.

Laws of nature that guide the natural mechanisms of the universe are manifested with probabilistic characters rather than deterministic ones. This was uncovered by quantum theory, one of the pillars of modern physics. This is also exemplified in other fields of science; the laws of chemistry and the life sciences which employ atomic and molecular reactions to explain biological phenomena. Therefore, the laws of biochemistry and ecological transformations do not hold deterministic results; on the contrary, these processes are based on probabilistic outcomes, a fact that has been proven by modern science and the discoveries of the twentieth century.

Once we recognise that biological operations have a large number of possibilities, it is reasonable for us to wonder about what controls these possibilities. It is necessary to realise that no single molecule of the reactants can be thought to assume either control of the driven operation or the choice of the probabilistic results. Such a 'controller' who dictates a specific choice of occurrence of an event with a low probability must be knowledgeable of all the parts and their properties, as well as with their conditions and requirements. In addition, the controller has to be supreme to all choices and purposes so that a positive evolutionary result is obtained. This echoes what the British physicist Paul Davies, quoting the famous astrophysicist Fred Hoyle, said: "A commonsense interpretation of the facts suggest that a super intellect has monkeyed with physics, as well as chemistry and biology and that there are no blind forces worth speaking about in nature."[63]

Does the Qur'an Oppose Biological Evolution?

The answer to this question is twofold: one concerns the evolution of man and the other concerns the evolution of other creatures. With regards to the latter, the Qur'an mentions nothing about the rest of the creatures on Earth, except telling us that:

> *And there is no animal in the Earth, nor a bird that flies with its two wings, but (they are) communities like yourselves. [6:38]*

{وَمَا مِنْ دَابَّةٍ فِي الْأَرْضِ وَلَا طَائِرٍ يَطِيرُ بِجَنَاحَيْهِ إِلَّا أُمَمٌ أَمْثَالُكُمْ} [الأنعام:38]

---
63 Davies, *The Accidental Universe*, 118.

Below, I will present my understanding of the Qur'anic texts with reference to creation, with the purpose of investigating the truth as presented by the Qur'an. The methodology of this research is based on reviewing the Qur'anic texts concerning creation and the emergence and evolution of man, whilst taking into consideration the basics of the Arabic language. Additionally, I take into account the many interpretations accredited to previous Muslim scholars, one of which is the interpretation of Ibn Kathir.

The main concern of the Qur'an in narrating the story of the creation and evolution of man is to specify that this event was ordered by Allah according to His will. We understand that Allah informed the angels of His will but the angels were confused. We read:

> *And when thy Lord said to the angels, I am going to place a ruler (a placement) in the Earth, they said: "Wilt Thou place in it such as makes mischief in it and sheds blood? And we celebrate Thy praises and extol Thy holiness." He said: "Surely I know what you know not." [2:30]*
>
> {وَإِذْ قَالَ رَبُّكَ لِلْمَلَائِكَةِ إِنِّي جَاعِلٌ فِي الْأَرْضِ خَلِيفَةً قَالُوا أَتَجْعَلُ فِيهَا مَن يُفْسِدُ فِيهَا وَيَسْفِكُ الدِّمَاءَ وَنَحْنُ نُسَبِّحُ بِحَمْدِكَ وَنُقَدِّسُ لَكَ قَالَ إِنِّي أَعْلَمُ مَا لَا تَعْلَمُونَ}
> [البقرة:30]

It is not clear why the angels questioned the will of God, but it is clear that they suspected this 'being' as someone who might engage in mischief and bloodshed. This indicates that they had prior experience with such a creature. This can only mean the available creatures at the time, i.e. animals and the like. In another verse, the Qur'an tells us:

> *(Remember) when your Lord said to the angels: Truly, I am going to create man from clay. So when I have made him complete and breathed into him of My Spirit, fall down submitting to him. And the angels submitted, all of them. [38:71–73]*

The Divine Word and the Grand Design

{إِذْ قَالَ رَبُّكَ لِلْمَلَائِكَةِ إِنِّي خَالِقٌ بَشَرًا مِنْ طِينٍ (71) فَإِذَا سَوَّيْتُهُ وَنَفَخْتُ فِيهِ مِنْ رُوحِي فَقَعُوا لَهُ سَاجِدِينَ (72) فَسَجَدَ الْمَلَائِكَةُ كُلُّهُمْ أَجْمَعُونَ}
[ص:71–73]

When taken in isolation this verse may give us the impression that God gathered some mud and moulded it into the form of man, then breathed into him of His Spirit so that he became a living being and thereafter the angels submitted to this new creature. However if we investigate other verses of the Qur'an concerning this topic, we see that there are staggering details that should not be overlooked. These details explain the stages by which man evolved to be a creature deserving the submission of the angels. The Qur'an tells us that God started creating man from clay:

> ...and He began the creation of man from clay. [32:7]

{الَّذِي أَحْسَنَ كُلَّ شَيْءٍ خَلَقَهُ وَبَدَأَ خَلْقَ الْإِنْسَانِ مِنْ طِينٍ} [السجدة:7]

But what sort of clay was this? The Qur'an tells us that it was wet. Regarding this, there are three verses in the same chapter Surah al-Hijr, where we read:

> And surely We created man of sounding clay, of black mud fashioned into shape. [15:26]

{وَلَقَدْ خَلَقْنَا الْإِنْسَانَ مِنْ صَلْصَالٍ مِنْ حَمَإٍ مَسْنُونٍ} [الحجر:26]

> And when thy Lord said to the angels: "I am going to create a mortal of sounding clay, of black mud fashioned into shape." [15:28]

{وَإِذْ قَالَ رَبُّكَ لِلْمَلَائِكَةِ إِنِّي خَالِقٌ بَشَرًا مِنْ صَلْصَالٍ مِنْ حَمَإٍ مَسْنُونٍ} [الحجر:28]

But the *Shaytaan* (the devil) objected saying:

> "I am not going to make obeisance to a mortal, whom Thou has created of sounding clay, of black mud fashioned into shape."
> [15:33]

{قَالَ لَمْ أَكُنْ لِأَسْجُدَ لِبَشَرٍ خَلَقْتَهُ مِنْ صَلْصَالٍ مِنْ حَمَإٍ مَسْنُونٍ} [الحجر:33]

In the above three verses we spot the words 'sounding clay' (according to some translations, 'sounding pottery') then the words 'black mud', which indicates a mud that is old, mineral-rich and decomposing. This understanding is well-supported by texts from traditional Qur'anic exegesis; therefore, we can consider it reliable enough to express the meaning of the verse. Accordingly, we can conclude that God began to create man from decomposing mud. But where can such mud be found, except in ponds? Incidentally, the exact words of the verse say that man was created from *hama' masnoon*—the word *hama'* meaning old mud while *masnoon* could have several meanings. One potential meaning is that it is old and has been there for many years, and the other, which I find more meaningful, is that it means old mud which has been assimilated for the intended purpose, i.e. the creation of man.

The question then arises as to whether the creation of other creatures was from the same mud or not. There is no indication in the Qur'an about this, but experience tells us that the style of the Qur'an when presenting a topic focuses on the main issue rather than the details. The topic here is the creation of man and there is nothing about other creatures being created in the same way. However, we should remember that we are dealing here only with the beginning, i.e. the first stage.

In the second stage of creation, the Qur'an tells us about the preparation and development of man. This we can understand from several verses, which mention the care taken in making man. We read:

> So when I have made him complete and breathed into him of My Spirit, fall down making obeisance to him. [15:29]

{فَإِذَا سَوَّيْتُهُ وَنَفَخْتُ فِيهِ مِنْ رُوحِي فَقَعُوا لَهُ سَاجِدِينَ} [الحجر:29]

The Divine Word and the Grand Design

The same verse is repeated in [38:72]. We also read:

> *He Who created you, fashioned you perfectly, and enabled you to be upright. [82:7]*

{الَّذِي خَلَقَكَ فَسَوَّاكَ فَعَدَلَكَ} [الانفطار: 7]

Many of the available translations of the Qur'an do not give the correct translation of this verse, perhaps out of lack of awareness or to avoid any indication of evolution in the ascent of man.

A narration of the Prophet Muhammad ﷺ was documented by Ibn Kathir, indicating that the fashioning stage correlated to the ability of man to walk upright on his feet, which is considered to be one of the stages of human evolution. The Prophet Muhammad ﷺ said: "God says: Oh the son of man, how would you challenge me while I have created you out of this [spit], and once I fashioned you upright you walked proudly collecting [wealth] meanly and once you feel dying, you say now I will be charitable and what a time for charity." This important hadith qudsi is widely considered to be authentic.[64]

The third stage of man's development was being imbued with the Spirit of God through the breath of Allah. At this moment, man became human through acquiring partial will and intellect. This enabled him to choose, invent, create and develop things for his benefit or, unknowingly, his misery. God then ordered the angels to prostrate themselves before the human being, indicating that they, along with the rest of the world, would be subservient to this creature. During this stage, man evolved to the level of a human being by obtaining cognitive abilities. The Qur'an tells us that at this point Allah taught the new creature, i.e. Adam, the names of things as an indication of having acquired cognitive capabilities:

> *And He taught Adam all the names, then presented them to the angels; He said: "Tell Me the names of those if you are right." [2:31]*

---

64 Ahmad Ibn Hanbal, *The Musnad*, Risala Edition, vol. 29, 385.

{وَعَلَّمَ آدَمَ الْأَسْمَاءَ كُلَّهَا ثُمَّ عَرَضَهُمْ عَلَى الْمَلَائِكَةِ فَقَالَ أَنْبِئُونِي بِأَسْمَاءِ هَٰؤُلَاءِ إِنْ كُنْتُمْ صَادِقِينَ} [البقرة:31]

The subsequent story of mankind, as told in the Qur'an, informs us that propagation of the species was accomplished through mating between the sperm of the males and the eggs of the females, referred to as 'fluids despised'. Here, we see in the Qur'an two verses which appear to have two different meanings, the first saying:

> *Then He made his offspring from semen of worthless water (male and female sexual discharge). [32:8]*

{ثُمَّ جَعَلَ نَسْلَهُ مِنْ سُلَالَةٍ مِنْ مَاءٍ مَهِينٍ} [السجدة:8]

While another verse states:

> *Did We not create you from a worthless water? [77:20]*

{أَلَمْ نَخْلُقْكُمْ مِنْ مَاءٍ مَهِينٍ} [المرسلات:20]

In my opinion, the first verse [32:8] addresses the creature that was developed prior to the formation of the first human being—Adam, so to speak. The evidence is the context itself, which intimates that the creature has not yet received the Holy Breath. Let us read the verses of Surah al-Sajdah in sequence. The verse is succeeded by the following:

> *... and He began the creation of man from clay Then He made his offspring from semen of worthless water (male and female sexual discharge). Then He fashioned him in due proportion, and breathed into him the soul (created by Allah for that person), and He gave you hearing (ears), sight (eyes) and hearts. Little is the thanks you give. [32:7–9]*

{الَّذِي أَحْسَنَ كُلَّ شَيْءٍ خَلَقَهُ وَبَدَأَ خَلْقَ الْإِنْسَانِ مِنْ طِينٍ (7) ثُمَّ جَعَلَ نَسْلَهُ مِنْ سُلَالَةٍ مِنْ مَاءٍ مَهِينٍ (8) ثُمَّ سَوَّاهُ وَنَفَخَ فِيهِ مِنْ رُوحِهِ وَجَعَلَ لَكُمُ السَّمْعَ وَالْأَبْصَارَ وَالْأَفْئِدَةَ قَلِيلًا مَا تَشْكُرُونَ} [السجدة:7-9]

Again, in order to avoid indicating that there are two (or more) different creatures among the origins of human descent, most exegeses of the Qur'an choose to interpret this verse to consider the Holy Breath to be an event which occurs at a certain stage of the foetus' development in the womb. However, this interpretation is in conflict with the phraseology of the verse which mentions the 'offspring' (*naslahu*), which is not a foetus but a newly-born child.

Furthermore, the breathing of the Holy Spirit into the body, as presented in the third stage of creating a human being, was a turning point in the evolution process. The creature was transformed from the uncivilised stage where he had no developed mind into a stage where he could think and deduce using an advanced intellect. This, I believe, was imparted by Allah through His Spirit. His breath altered man from being an upper animal into a human being. Following this stage, the angels (which here is a metaphor standing for the laws of nature) were ordered to be at the disposal of humankind, to enable him to explore the world, discover it and exploit it positively for his own welfare. Through this, man was supposed to acknowledge the greatness of his Creator, glorifying Him in order to become part of His high-ranking kingdom.

From the interpretation given by Ibn Faris in his lexicon of Arabic,[65] in which he provides the origin of words out of their literal construct, I understand that the bowing of the angels before man was intended to put them at the disposal of humankind. Of the word bow (*sajada*), he says that its origins point to lowness and submissiveness. It is commonly known that a low-ranked subject prostrates before a higher-ranking person. The reason for the angels bowing revolves around the fact that they are Allah's agents, His messengers to the world, and those through whom Allah orders the world. In other words, the angels prostrated before Adam because he was given priority and privilege above them, and superiority due to his strength and deeds.

---

65 Ibn Faris, *Mujam Maqayis al-Lugha*, edited by Abdul Salam Harun, Dar al-Fikr, Beruit, 1979. Entry: sajada.

Based on this presentation and analysis of Qur'anic verses, it may be concluded that the creation and development of humankind does not necessarily contradict with the possibility of biological evolution through which humankind has developed to its present stage. In terms of creation and evolution, it is emphasised in the Holy Qur'an that all that happens is predestined by Allah and takes place according to His will and the laws of logic. Additionally, there is a possibility for having an evolutionary setback or a developmental change resulting in a malformed creature; in other words, a metamorphosis.

Here I am not claiming that our understanding of the Qur'anic image is a comprehensive one; on the contrary, there must be an unknown, metaphysical aspect to the image which involves the creation of Adam in the Garden. It cannot be a clear-cut conclusion that the Garden is on Earth, although some scholars consider that the issue of being removed from the Garden and sent down to Earth was only meant to serve a spiritual purpose. Ultimately, the unintelligible passages in the Holy Qur'an are in need of further analysis in order to devise solutions and decipher their mysteries.

Another significant point which might be thought to conflict with religious belief is the claim that man originated as an animal: a monkey, a deer, a fish or some other creature. Some religious doctrines (not necessarily the Qur'an) confirm the idea that these creatures were the result of metamorphoses which do not conform to the concept of honouring man and making him a vicegerent on Earth by the Creator. However, this problem does not arise in the Qur'an.

The honouring of mankind as the sons of Adam arose after the advanced creature became Adam-like. The primitive creature became a human being when the Creator fashioned him, straightened him and blew the Holy Spirit into him. The morphological, anatomical, physiological, and even behavioural facts indicate a similarity between the species, known as homo sapiens, and the other advanced creatures. We as humans, eat, drink, breed and practise many of our daily life activities in a way that resembles those of other animals. The difference between us and other high-ranking animals is simply that we have the skill and constructive logic by which we are able to comprehend the world. Without this talent, we might descend into the lower animal class. Should we choose to ignore the fact that not only is there a purpose behind our existence, but there is also a power, a will and an objective for the world in which we live, then we will resemble other members of the animal kingdom to which we originally belonged. Put simply, our class is spiritually advanced compared to the animal kingdom inasmuch as we have the ability to distinguish, innovate and construct. Yet if we underestimate this gift, along with the facts to which it guides us, we will become like cattle or even worse. Unlike humans, animals do not have rational faith. The Creator raised man over other creatures to enable him to think, in order to be the vicegerent who innovates and perceives the universe around him using his senses (albeit only a small portion of the physical part of the universe). Accordingly, a human being has to appreciate these values and respect them at all levels and in all activities. Man and the creatures of the animal kingdom are equal except for what the former can acquire of these superior qualities.

Scientific Analysis

The notion of biological evolution of living creatures is a reasonable explanation for numerous observational facts, which is what makes it a fundamental aspect of modern life sciences. It would be difficult to recognise modern biology as a major field of science if it was devoid of this principle. Those who deny biological evolution should submit a legitimate rationalization for the emergence of species on Earth. This should also be capable of justifying the morphological, physiological, and structural similarities, in addition to the innate sociological interaction between themselves and other creatures of the animal kingdom. Moreover, they would have to explain the success of those scientists who adopted biological evolution in their research, as this success indicates a correct approach that has been followed over many decades.

However, we should be aware that biological evolution does not necessarily imply the endorsement of the neo-Darwinian approach for evolution. It does not mean that all theoretical explanations proposed by the neo-Darwinians are correct, as scientific evidence suggests that some theoretical details are still unknown. Consequently, a distinction should be drawn between acknowledging biological evolution on the one hand, and believing in the theories of evolution on the other. The former demonstrates an established reality which is supported by a great deal of evidence, whereas the latter is controversial and should not be taken for granted.

Some authors such as Jean Staune[66] consider the theory of evolution as being incomplete and observe that it is somewhat similar to the theories of motion set during the times of Galileo, before the discovery of Newton's laws. This suggests that the current theory of biological evolution needs to undergo many transformations in order to proffer the reality of evolution as a complete sequence. For example, the claim adopted by the neo-Darwinists that evolution rests on two pillars—random mutations and natural selection—is incompatible with the fruitful results of evolution that we see occurring in nature.

Darwinism purports that the division of sexual cells results in random mutations within the genetic material. Consequently these mutations, which emerge during the copying process, cause changes in the hereditary code and their effects may become manifest during the development of the organism. During its lifetime, a conflict for survival is perpetually taking place between the creature and its natural habitat. Only when the result of this conflict is known will it be shown whether or not this biological mutation is of benefit. If the mutation is a beneficial one, this will lead to the survival of the creature and give it a better opportunity of breeding than those with DNA that did not incorporate

---

66 Staune, J. *Does our existence have a meaning?* Paris: Presses de la Renaissance, 2007.

the mutation. On the other hand, if the mutation results in a severe impairment then the creature may die, as might other organisms with equally non-favourable mutations. Since nature dictates whether or not creatures with developmental mutations are to survive and reproduce, the selection process has been bestowed the term 'natural selection'. Logically, the creatures carrying the preferred mutations will breed and be prosperous while those carrying detrimental mutations will decrease and become extinct. This is an illustration for the mechanism suggested by Darwin's theory of evolution, according to the modern understanding.

Biological evolution, as I have shown, is acceptable and can be easily accommodated within the framework of the Islamic worldview, provided an accurate interpretation of the Qur'an is made available. In fact, there is no major inconsistency between Darwinism and Islamic belief, except when assuming genetic mutations to be totally random. If one believes that mutations are entirely random, then it implies that they are independent of the Creator's will. The fact that the results of these laws are based on probability rather than deterministically negates the notion of randomness and asserts the principle of indeterminism.

Necessity and Contingency in Evolution

There are some significant loopholes in the neo-Darwinian explanation of evolution and alternative non-Darwinian approaches have been suggested.[67] The purpose of this section is not to delve into the details of such criticism, but to stress again that the random mutation/natural selection process cannot be accepted as a mechanism for generating new, fruitful and ascending evolution as observed in nature. It cannot guarantee the efficient diversity of creatures on Earth. Random mutations mean equally probable events that may over a long period of time cancel out and produce nothing, despite the alleged natural selection factor.

---

67 King, Jack Lester and Jukes, Thomas H, "Non-Darwinian Evolution" *Science* 164, 788–798.

The mechanism of evolution might be understood through the existence of two necessities and many contingencies. The first necessity is the need to adapt to environmental conditions. For this purpose, new features in the morphology or the physiology of the creature are essential. On the other hand, the blind law of nature necessitates both a driver and a coordinator in order to guarantee a harmonious and fruitful output. For each of these necessities there are many contingencies—and perhaps endless possibilities—but there will always be only one optimised solution. Such a solution picks those contingencies which will ultimately achieve maximum efficiency and harmony. However, since the laws of nature are probabilistic, there will always be a few choices of un-optimised contingencies. This could potentially result in failure, with evolution producing undesired traits. This vision may provide a basis for a mechanism of evolution that satisfies better observational facts.

Microscopic Evolution on a Quantum Base

The behaviour of atoms and molecules in chromosomes modulate biochemical reactions, which may result in genetic mutations. The laws of quantum mechanics control this behaviour microscopically. They were formulated in the first half of the 20th century as a result of the research of Max Planck, Niels Bohr, Werner Heisenberg, Louis de Broglie, Erwin Schrodinger and many others. The laws had a logic regarding the microscopic behaviour of particles, which was somewhat different from the traditional logic of classical physics. The most important features of this logic were related to probability and determinism; events that happen are probabilistic, i.e. they are singled out of many contingent events that depend on the state of the system in conjunction with the process of measurement. The basic aspects of this theory have been clarified in chapter three of my book *God, Nature and the Cause*, especially the issue of measurement of quantum mechanics, along with different explanations for this theory. I have contributed my own explanation for

The Divine Word and the Grand Design

the problem of quantum measurement which is based on the notion of re-creation, as borrowed from daqiq al-kalam. Aside from the chronic problem of measurement, quantum theory is the best available framework for understanding the interaction of matter and energy, and the behaviour of elementary particles, atoms and molecules. Since biochemistry is concerned with reactions between these atoms, molecules and ion transport to form different compounds (the components of biological materials of the living substances), quantum mechanics and logic are undoubtedly linked to understanding these compounds through the topic of quantum biochemistry.

Johnjoe McFadden, a molecular genetics professor at the University of Surrey, attempted to make use of quantum mechanics and biochemistry in order to work out chemical evolution at the molecular level of basic biological compounds in a proto-cell. He presented the general framework of his theory in a book entitled *Quantum Evolution: Life in the Multiverse*,[68] in which he states that these quantum activities have an effect on the whole structure of a living creature due to the fact that a single DNA molecule may affect all the other cells. Since microscopic composition of the molecules is ruled by quantum indeterminism, quantum mechanics undoubtedly has an essential role in composing the amino acids and the alkaline bases that constitute genes—the hereditary elements of living creatures.

It is also known that the possible number of protein compounds which can be made up from these amino acids and alkaline bases can run to billions; whereas, in reality, we see only a small number of these compounds formed. This is a real problem faced by biochemistry at the molecular level. In order to clarify the size and importance of this problem, McFadden undertook the construction of peptides as a result of the interaction between 20 alkaline bases and 32 amino acids. The number of probable compounds based on random distribution was estimated to be 2032 which means 1041 compounds—a very large number. If

---

68 Johnjoe McFadden, *Quantum Evolution: Life in the Multiverse,*, HarperCollins, 2000.

Evolution

we hypothesise that the random constructions are going to make only one molecule of each of the peptides, then this will require a lot of carbon, much more than is available in all the trees on Earth. Ideally, there should be a rule which governs these probable selections and greatly decreases the available number of peptide chains. For this purpose, McFadden utilised the superposition of states principle, a fundamental pillar of quantum mechanism postulated by the Copenhagen School. This explanation maintains that the quantum physical system, in the absence of measurement, is a superposition of all possible states. At the moment of measurement the physical state of a system collapses on one of the possible states (in what is called 'wave-function collapse'), or the symmetry between the observer and what was assigned on the scene as previously mentioned. McFadden devoted a great deal of effort to justifying the availability of the quantum states in limited peptides chains, which can breed automatically. He explains this descriptively without providing any detailed calculations.

However, McFadden encounters a major problem in terms of quantum measurement when he asks: "Where have other peptide chains disappeared?"[69] Here, McFadden employs the idea of a multiverse, hypothesizing the existence of 2032 worlds to make our universe the only one which acquires the peptide that is capable of breeding automatically. But McFadden immediately understands that the multiverse idea necessitates the notion that what has happened in one universe will not happen in other universes. Were it applicable to the universe in which we live, it would be a unique event which could not be duplicated. In other words, McFadden suggests the scenario to be one that cannot reoccur in any other place in our universe. This means that life in our universe can only be expected to be on Earth. Based on the same argument from the multiverse interplay, McFadden concludes that it is impossible to repeat the development of life in a laboratory experiment. He believes that even if scholars gather all required conditions to establish life in a laboratory,

---

69 McFadden, *Quantum Evolution*, 227.

they will not be able to regenerate life. This harks back to what the physicist Lee Smolin said regarding the multiverse hypothesis—that it can neither be proved, nor disproved.

In summary, it is clear that McFadden's reliance on the Copenhagen interpretation of quantum measurements, in addition to exploiting the notion of the multiverse in his attempt to present a scenario on producing self-replicating peptides, is not adequate enough to prove the validity of the form. There were additionally no detailed calculations in this regard. Matthew Donald of the Cavendish Laboratory in Britain wrote an essay criticising what McFadden presented and describing his method as completely wrong.[70] McFadden and Al-Khalili responded to Donald's essay, providing arguments in favour of their scenario. In general, what McFadden suggested on the process of producing self-replicating peptides is still incomplete, although it contains some interesting points in proposing a major role for quantum theory on the development of the creature in its early stages.

Difficulties with Neo-Darwinism

As mentioned before, neo-Darwinism is based on two principles: random mutations and natural selection. The aim at this point is not to discuss the theory of evolution in detail but rather to focus on the general, as well as the fundamental, aspects of this theory. The discussion is supported by a number of general proofs which are characterised by negating false information through verifying the inconsistency in results aside from the procedural details. We may consider that natural selection is actually possible, whereas hypothesizing about random mutations is more problematic due to the fact that randomness cannot escalate fruitfully, as is the case with the evolution of living creatures.

---

70 Donald, M. J. Book Review of Quantum Evolution by Mathew J. Donald, Jan 2001, arXiv. quant-ph/0101019.

Advocates of Darwinian evolution believe that the advancement of living creatures is a result of cooperation between natural selection and useful mutations. However, a simple calculation involving the random options required for the development of a complex organ would reveal the need for a great number of eras (perhaps longer than the age of the whole universe) in order to create a sizeable qualitative development in those organs. This problem may be overcome if there is a steered development; that is, mutations which are not random but somehow guided by factors from within or without the creature through an as yet unknown power.

To say that mutations occur at the behest of some unknown actor might imply that the universe is run by miracles. This, in turn, would invalidate causality and deprive science of the most efficient element for its credibility. Such an invalidation of causal relationships is not acceptable in the new kalam. Causal relationships are acknowledged, secondary causes exist, but causal determinism is refuted. Mutations are undoubtedly a fact of life that can be verified in a laboratory, but to explain them as occurring randomly involves metaphysical assumptions. It is saying implicitly that for some unknown reason, mutations are occurring. Describing mutations as 'a mistake' in copying the DNA is again not an adequate portrayal of what could actually be happening. For example, psychological issues such as stress may sometimes play a role in producing genetic mutations, while variations in the Earth's magnetic field may have substantial effects in directing the tiny ions involved in the genetic reproduction process. There are many pieces of evidence on physical-psychological effects from various perspectives, such as sudden fright and how it can cause physiological syndromes due to the quick and excessive production of certain hormones.

An important piece of experimental work was conducted by the American biologist John Cairns, the findings of which were published in 1988.[71] Cairns isolated a group of bacteria unable to utilise lactose

---

71 Cairns, *Nature*, 1988.

and grew them on a nutrient-poor medium such that their numbers decreased dramatically. When he transferred several of these clones onto a lactose-rich medium, he observed that some of the bacterial clones were able to grow and produce colonies. These, he subsequently discovered, had acquired genes that enabled them to metabolise lactose and consequently survive. The findings of this experiment were considered controversial because there were no known mechanisms explaining how genetic development could be directed in such a way that it would produce the genes necessary for the breakdown of sugars. The known evolutionary mechanism of genetic mutations stated that the route of information was from DNA through the RNA to the protein, and not vice versa. Until recently, there has been no known biological mechanism which enables the movement of information from the environment and surroundings towards the genes to urge its division. This is the problem.

Johnjoe McFadden published a paper in collaboration with Jim al-Khalili which considered adaptive genetic mutations in terms of quantum mechanics.[72] Although after 1988 many research papers were published which supported the idea of the occurrence of adaptive mutations, they did not provide a viable explanation. However, does this lack of explanation for certain phenomena give us an excuse to reject it?

The advocates of Darwin's evolutionary theory emphasise that all life originated from one cell and therefore, that all the species of all living creatures have one single point of origin. This is the reason why the length of time required to bring about these variations in species is under scrutiny and not decisively definite. However, it may be speculated that it might be a very long time, extending perhaps beyond the age of the Earth. This is another argument against the claim for random mutations.

Let us take another example. Suppose we have a large number of similar tools made of the same material but in different shapes (household cutlery, for example) and we ask someone to work out the relationship

---

[72] McFadden, J. and al-Khalili, J. A Quantum Mechanical Model of Adaptive Mutation, *BioSystems* 50 (1999) 203–211.

between these tools. He might say they are made by one person, and if not by one person, then at least the source of knowledge is common for all manufacturers. If we tell him that we do not have convincing evidence for the existence of this one manufacturer or those common knowledge manufacturers, he might conclude that these tools may well have made themselves. Then we might ask him whether each tool makes another one or are the tools generated from each other? Of course, the simplest reply will be that they generate spontaneously; they must have developed from each other according to the needs they encountered. If we tell him that these tools do not have intelligence, no hindsight to distinguish or a will, and that they are blindly copying each other, he might say: Aha! They must have developed randomly—a knife changed into a spoon and the spoon into a fork, for instance. If the knife randomly changes into an impractical object, then it will be discarded since it will have no use. Based on this argument, the same person might say that these devices must have been used for different purposes and that the purpose of the device determines its material and shape. Accordingly, external factors decide the shape of the devices. If the same person finds one of the metal tools containing a plastic part, he will be surprised and ask about the source of plastic. He will question the feasibility of the metal changing into plastic. He will become more and more perplexed but maintain that these tools result from gradual and slow changes from one and the same origin.

This analogy is similar to the arguments put forward by Darwinists from material propagated by Richard Dawkins, who denies the existence of God and does not allow himself to consider the possibility of a creator who can direct evolution through certain laws and choose among the probabilities of the possible mutations. Who can falsify the theory of evolving cutlery?

A Summary of Biological Evolution

The common belief that the evolution of living creatures contradicts with religious convictions originates from the concerns of some people that accepting evolution involves denying the role of the Creator. They believe that the laws of nature would then automatically replace the will and act of God. However, people should realise that this is a false notion, since the effectiveness of the laws of nature is limited by the probabilities of the results and these probabilities are not chosen by the laws themselves. Moreover, such a fact uncovers that the laws of nature themselves, including the laws of biological evolution, are in need of a driver that cannot be part of the constituents of the universe.

This does not necessarily mean that the Holy Qur'an supports the evolutionary theory of Darwin or any other. The *principle* of evolution is different from the *theory* of evolution. This is because the theory at hand is replete with deficiencies, loopholes and draw-backs. The idea of the emerging species based on the assumption of completely random mutation cannot be accepted. Evolution is ascending and it is difficult to ascertain how such a well-organised hierarchy can emerge out of random chances.

In short, the *theory* of evolution is distinct from the *principle* of evolution. In my view, the Holy Qur'an does not object to the notion of the biological evolution of human beings, nor does it set out any specific details for the mechanisms of this evolution and its various stages. On this occasion it is necessary to emphasise once again that those who reject the biological evolution of man have not yet accounted for many of the examples presented by biological events, with reference to the biological development of living creatures. Nor have they provided an explanation for those verses in the Qur'an that address the different stages of creation, especially those in Surah al-Sajdah. Any elucidations they might provide should meet the high standards of the Qur'an, with its accurate vocabulary and its precise expressions.

Chapter Eight

The Universe

The universe encompasses everything we can see when we look beyond the Earth's surface. It is the vast space that extends farther than what can be seen with the naked eye. During the Greek and Islamic civilisations, the term used to describe this was the 'world'.

The Creation of the Universe

For millennia, people thought that the universe had no beginning and that it would remain in its current state for eternity. They did not even realise that it extends beyond what they called the 'Sphere of Fixed Stars'. Ancient natural philosophers such as Plato and Aristotle believed that the Earth was the centre of the world, which they divided into two parts: one was the lunar sphere and beyond, considered to be the immutable ethereal world, and the second was the Earth and its atmosphere, the mutable world of corruption and change. The Earth, at the centre of the world, was surrounded by seven spheres listed in ascending order: the lunar sphere of Mercury, followed by Venus, the Sun, Mars, Jupiter and Saturn, beyond which lies the eighth sphere of the fixed stars. This is the geocentric system of the world. Most Muslim philosophers

The Divine Word and the Grand Design

and astronomers adopted this picture of the world and most commentators of the Qur'an expressed their understanding of verses in line with it.

Ikhwan al-Safa, a religious group from the Ismaili sect, described this system beautifully in their *Letters of Ikhwan al-Safa*. They followed the general scheme of the ancient astronomers in placing the Earth at the centre followed by the seven planets: Moon, Mercury, Venus, Sun, Mars Jupiter, and Saturn, then the fixed stars and finally the outermost sphere, or the *Muhīt*. This was added by Muslim astronomers to the spheres of Ptolemy to account for the precession of the equinox. Thus, in total they proposed that there were nine spheres.

Using the geocentric model of the world, Ikhwan al-Safa provided an explanation of the seven Heavens and the seven Earths mentioned in verse [65:12] where it is stated that the seven heavens are the seven spheres, beginning with the lunar sphere and ending with the sphere of Saturn. These are arranged consecutively in concentric layers such that each one is a heaven for the lower one and is a land (Earth) for the upper sphere above it. Moreover, they equate the eighth heaven of the fixed stars with the *kursi* (pedestal) mentioned in the verse, and the ninth heaven with the *'arsh* (throne) by interpreting the verse [69:17] on the basis that the lower eight spheres are carrying the 'arsh in accordance with Qur'anic cosmology.

The Ptolemaic model was the guide for Muslim astronomers in conceiving the celestial construction. However, they also made a vast contribution to the criticism of this construction which helped Copernicus (1473–1543) to suggest the radically different configuration of the heliocentric system, in which the Sun is placed at the centre and around which the planets, including the Earth, rotate. Astronomers of the Muragha school, led by Nasir al-Din al-Tusi, Ibn al-Shatir al-Dimashqi and others, devised several theoretical models in order to interpret the observations which do not align with the Ptolemaic system.

The Universe

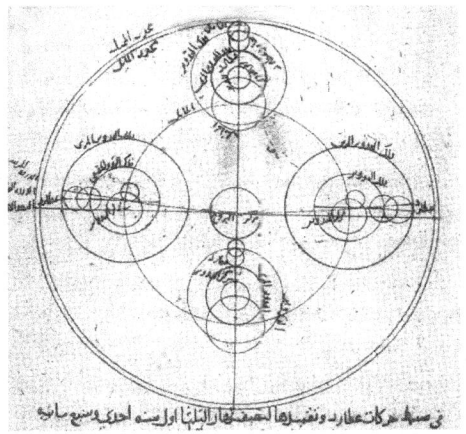

Fig. 14 Mercury orbits according to Ibn al-Shatir

Commentators of the Qur'an adopted the same view of the geocentric world, although some older commentaries contained stories and legends about the formation of the Earth and the heavens. As we can see from the commentary of Fakhr al-Din al-Razi (1150–1210) entitled *Mafateh al-Ghayb*—which was considered the most advanced elaboration of the scientific picture of the world at the time—most of the commentators of the Qur'an did not pay much attention to what the verses point to. He and other commentators were instead very much occupied with Aristotelian cosmology and its derivatives.

Undoubtedly, commentators of the Qur'an are affected by their beliefs and many carry a bias toward their knowledge and culture. Nevertheless, one should keep an open mind regarding other possible meanings and implications of the verses in order to maintain the real value contained in the Qur'an.

It is true that we see the stars, the Sun and the Moon rotating around the Earth and this is because we, the observers, are located on the Earth while it is rotating around the Sun. You can compare this state with a person riding a merry-go-round watching the motion of a ball thrown upward by another merry-go-round rider. The motion will appear distorted on the merry-go-round whereas it can be seen clearly when looked at on fixed land.

The heliocentric model makes life easier for the astronomer. You may compare the simple elliptic orbit of Mercury according to the heliocentric model with the complicated configuration of the planet orbs, devised according to the geocentric model in Figure 15.

Some ask why we should consider the Earth and the planets to be rotating around the Sun, while the Sun itself is moving in space at great speed?

Fig. 15 Motion of the planets in space viewed from the centre of the galaxy

Modern cosmology

By the beginning of the second decade of the 20th century, astronomers became confident that the universe is not just one island of stars but that there are many others. Each of these islands, known as galaxies, hosts billions of stars. Edwin Hubble was able to recognise dozens of these galaxies and tried to classify them according to their shapes. He also studied their velocities and found that most galaxies are receding from us with velocities proportional to their distance from us. The more distant the galaxy is, the higher its velocity is. This is called Hubble's law. It was found that only a few nearby galaxies were approaching us whereas most galaxies were receding from us, as if we were at the centre of the universe.

It is remarkable how man can measure the velocity of a distant object and know whether it is approaching or receding from us. There is no messenger between us and the distant celestial object except light. It was discovered that if a light source is receding from us, then the wavelength of its spectrum will be shifted to the red end, and if it is approaching us, then the light emitted by that object will be shifted to the blue end of the spectrum.

The relative amount of shift in the wavelength is a measure of the velocity of the object. By analysing the light received from different galaxies, by the end of the second decade of the 20th century Hubble was able to declare that the universe as a whole is expanding. This opened up a new era in the history of cosmology. His discovery radically changed the picture of the universe from being considered static to dynamic.

Hubble's discovery also made an impact in the wider scientific community. Albert Einstein changed his views about the universe (which he had assumed to be static) and as a result, modified his original equations which produced a collapsing or expanding universe by adding a repulsive force that can balance the collapse. He called this force the cosmological constant.

At the same time in Russia, Alexander Friedmann was solving the Einstein field equation without introducing any cosmological constant. He obtained three possible solutions, as shown in Figure 16.

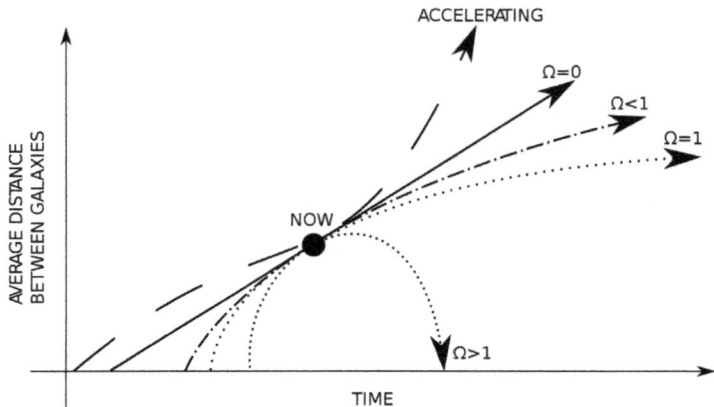

Fig. 16 The three possible solutions of the Friedmann equations

The first solution is an open universe which expands with acceleration throughout history. Geometrically, this universe has negative curvature. A surface with negative curvature is like a saddle; if we draw a triangle on its surface, the sum of its angle would be less than 180 degrees. It would go on expanding for eternity. The second solution is a flat universe, meaning it is like a sheet of paper on which we can draw a triangle with the total sum of its angles, 180 degrees. This universe starts accelerating but reaches an ultimate constant velocity. The third solution is a closed universe, with positive spatial curvature like the surface of a sphere. If we draw a triangle on such a surface, its angles add up to more than 180 degrees.

All of the three mathematical solutions proposed by Friedmann start from a point singularity; that is, a point with zero dimensions at time $t=0$. This means that the universe began from nothing. All the matter/energy content of the universe sprang into existence all of a sudden from nowhere. Some authors refer to the cosmic singularity as a point with infinite density, but this is not accurate enough.

The Universe

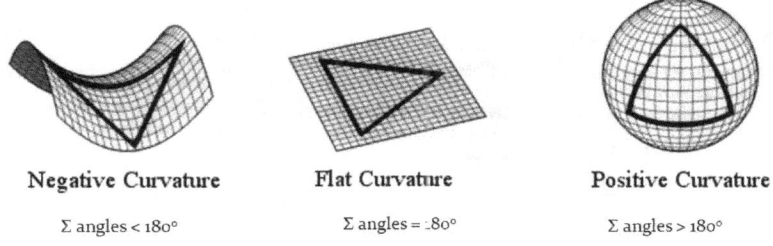

Negative Curvature
Σ angles < 180°

Flat Curvature
Σ angles = 180°

Positive Curvature
Σ angles > 180°

Fig. 17 Two-dimensional analogy showing the curvature of the spatial part of the universe.

Physically, in an open universe, the average density of matter is less than the prescribed amount called critical density. The gravity of such a density is less than enough to hold the universe from expanding at a fixed rate, hence, the universe goes on expanding with acceleration.

In a flat universe, the average density is equal to the critical density. This amount will enable gravity to hold the universe and ultimately attain a steady rate of expansion; nevertheless, it will continue expanding indefinitely. In a closed universe, the average density is higher than the critical density and such a universe will expand with deceleration until it reaches a maximum size, at which point the expansion turns into contraction.

The Expanding Universe

Why and how is the universe expanding? Why do we observe all other galaxies receding from us? Does the expansion of the universe cause the distance between the Sun and the Earth or the Moon and the Earth to increase?

The universe is expanding because it has been created to develop in time. The driving force behind this expansion has been labelled 'dark energy'. No one is yet certain of what this dark energy is. In my own

PhD work, I calculated the energy that is produced when the quantum vacuum is confined into a finite space and found that there is some positive energy created called 'Casimir energy' due to the confinement of the quantum vacuum.[73] This energy could back-react at finite temperatures on the geometry of the universe through the Einstein field equations, producing expansion of spacetime.[74]

The Qur'an alludes to the expansion of the universe, where it states:

> *We constructed the sky with our hands, and we will continue to expand it. [51:47]*

{وَالسَّمَاءَ بَنَيْنَاهَا بِأَيْدٍ وَإِنَّا لَمُوسِعُونَ} [الذاريات:47]

This statement contrasts with the general belief at the time of revelation, which corresponded to the Aristotelian idea that the heaven has a fixed size which cannot be increased or decreased, as it does not allow for generation or corruption.

The Big Bang theory

It was perhaps Georges Lemaître who first recognised the implications of an expanding universe. The Belgian physicist and priest rationally concluded that the universe must have had a beginning and could not be eternal. Accordingly, he declared that the universe may have been created as a small-sized object (which he termed the 'cosmic egg'), sprang out of the vacuum, and after a very short time began to increase in size. For this reason, some authors attribute the idea of the Big Bang theory to Lemaître, but historical facts tell us that Lemaître only identified the universe to have a beginning in space and time, contrary to the prevailing idea of his day. The term 'Big Bang' was in fact later coined by the British astrophysicist Fred Hoyle, and it was George Gamow and his

---

73 J.S. Dowker and M.B. Altaie, "Spinor Fields in an Einstein Universe: Vacuum Energy", *Phys. Rev.* D17, 417 (1978).

74 M.B. Altaie and J.S. Dowker, "Spinor Fields in an Einstein Universe: finite temperature corrections", *Phys. Rev.* D18, 3557 (1978).

collaborators, Ralph Alpher and Robert Herman, who established the Big Bang theory in its fully-fledged form with mathematical descriptions and calculations. This was completed in the late 1940s when these physicists tried to explain the natural abundance of chemical elements in the universe. Their observations showed that hydrogen, the most abundant element in the universe, forms about 76% of the material content, helium follows next at 23%, while the rest of the elements form no more than 1% of the material content of the universe. How could this be so?

Gamow and his collaborators suggested a scenario for the evolution of the universe starting with a very hot 'cosmic soup' that contains all the known elementary particles and their antiparticles in a state of thermal equilibrium. By thermal equilibrium we mean that these particles and their anti-particles were being created out of the available energy and immediately being annihilated (converted into energy again). As the universe was expanding in size, the steadily dropping temperature allowed for the heavy particles to go out of thermal equilibrium, i.e. they were no longer in the process of being created and immediately annihilated. The drop of temperature continued and throughout time, changes in the content of the universe occurred. Light nuclei of deuterium, tritium, helium and lithium were formed.

Formation of the first atoms

When the temperature reached about 5000 Kelvin, atoms were formed by combining the free electrons with these nuclei. This occurred when the universe was approximately 380,000 years old. Consequently, every atom released a small amount of energy. As the electrons and the nuclei combined together, the universe became transparent and the photons which were released by the newly formed atom could propagate through the space freely without obstruction. In other words, the mean free path became very large.

The Divine Word and the Grand Design

The released photons formed a heat bath that kept the universe warm for some time but as the universe expanded further, the temperature dropped further to about 3 degrees Kelvin, according to the latest calculation of the Gamow scenario.

Gamow and his collaborators explained the natural abundance of hydrogen, helium and lithium, but they could not explain how the heavier elements were formed.

Fred Hoyle and his collaborators later found that elements heavier than lithium are formed through complicated processes of nuclear reactions taking place inside stars. Sun-like stars can produce carbon and larger stars can form neon, oxygen, argon, silicon etc. up until iron. Elements heavier than iron up until uranium are formed in supernova explosions.

Cosmic Microwave Background Radiation

The distribution of the density of the cosmic microwave background radiation pattern in space contains information about the status of the universe in its very early stages. For this reason, I call this pattern the 'birth certificate' of the universe. As previously mentioned, formation of the first atoms was accompanied by the release of photons which had increased wavelength as the universe was expanding. This radiation formed the cosmic microwave background which can today be detected by microwave receivers, and is thus called Cosmic Microwave Background Radiation (CMBR). The radiation was accidentally discovered in 1965 by two engineers, Arno Penzias and Robert Wilson, while searching for extragalactic sources of microwaves.

Cosmologists analysed the spectral distribution of these radiations and found that it contained much information concerning the very early stages of the developing universe. This wealth of information is still growing and will certainly aid our efforts to gain a clearer understanding of the origin of the universe.

Some analysis of CMBR from 2002 unveiled that there were acoustic waves within the activities of the universe in its initial stages. I call such acoustic waves the 'Bang of the Creation'.

The Fate of the Universe

Modern cosmology tells us that the universe could have one of three possible fates. It may go on expanding forever in the case of an open universe, thus practically turning into empty space. In the case of a closed universe, it may collapse through a big crunch to turn into a fatal singularity, or potentially, it could collapse into a very small size in the far future before bouncing again into a new creation. This is called the oscillatory universe.

Practically, the current observational evidence tells us that the universe is accelerating and some cosmologists maintain that it will continue to expand forever. However, other evidence points towards the universe going into a collapsing phase since the curvature is not exactly zero, although very small.

The Qur'an tells us frankly that the universe will collapse on doomsday in a big crunch that will return it to the state in which it began. We read:

> *On the day when We roll up the heavens as if it were a written scroll and bring it back into existence just as though We had created it for the first time. This is what We have promised and We have always been true to Our promise. [21:104]*

{يَوْمَ نَطْوِي السَّمَاءَ كَطَيِّ السِّجِلِّ لِلْكُتُبِ كَمَا بَدَأْنَا أَوَّلَ خَلْقٍ نُعِيدُهُ وَعْدًا عَلَيْنَا إِنَّا كُنَّا فَاعِلِينَ} [الأنبياء:104]

This could be understood to mean that the universe will be of the oscillatory type. However, this model is the least expected by cosmologists of our time, due to several inherent flaws with it. Nevertheless, the above verse is very clear on the issue of the collapse; whether it will bounce into a new creation or not remains debatable.

The problem from a scientific point of view has been investigated by one of my postgraduate students who took the observational evidence at face value, thus assuming that the universe has been spatially flat from very early times and that there should be a non-zero cosmological constant within it. The cosmological constant is a repulsive force acting to make the universe expand. Accordingly, we devised a model taking a cylindrical geometry in four dimensions (three spatial dimensions and one time) with a time-dependent cosmological parameter. The Einstein field equations were solved and the results show that for such a universe there is the possibility of collapse, once the value of the cosmological parameter reaches a certain value. However, we also found that such a universe will bounce again to form a new creation through a new 'Big Bang' phase.

Since we are considering a three-dimentional flat surface, which we have taken as the surface of a cylinder embedded in a 4-dimentional space, we are actually representing the universe as a sheet. When the sheet collapses we find that it rolls up like a scroll in exactly the same way as mentioned in the above verse. A two-dimensional analogy of the collapsing flat universe is shown below.

Fig. 18 Two-dimensional analogy of the collapsing flat universe

The Gigantic Black Hole

As explained earlier, a black hole is a region of space with very strong gravity from which even light cannot escape. This region is surrounded by a spherical surface called the event horizon beyond which nothing is known.

Black holes are formed when a large mass (more than 3 times the mass of the Sun) collapses into a very small region of space. Physicists do not know any force that can stop the collapse at a finite size, which is the reason for believing that the collapse of the mass may continue to a point singularity. However, in practice, due to the quantum effects the collapsing mass might occupy a finite size. The black hole was named as such because it does not reflect any light; hence, it cannot be seen. Furthermore, time stops at the event horizon of the black hole and for this reason it is considered a hole in spacetime. It has been shown mathematically that inside the event horizon is a world in which time behaves differently. This region inside the black hole is said to be 'space-like'.

There is another speculative suggestion in cosmology which claims that if the universe is to collapse into a big crunch, then it will inevitably form a gigantic black hole. This black hole will sum up all the content of our universe. According to modern cosmology, this would be formed at the last stage of the collapse of the universe. But does this mean that everything in this universe will vanish into a worthless point? No! According to current scientific knowledge, no information in this universe will be lost if it becomes a black hole. For a long time Stephen Hawking, the famous black hole physicist, was of the opinion that information will become lost once an object turns into a black hole, but he later retreated from this opinion.

In his theoretical investigation of the information problem and black holes, the famous physicist Leonard Susskind found that the information going into a black hole will not vanish but will be exposed in holographic presentation on the surface of the event horizon. As proposed, by then all the acts of those who lived in this world—good and bad—will be exposed as three-dimensional images on the surface of the gigantic black hole. According to the Qur'an, the surface will resemble a huge screen showing what we have all done throughout our lives, precisely counted, person by person! This scrolled sheet collapsing into a line singularity and exposing information on its event horizon is the 'book' mentioned in the Qur'an:

> *And the book is placed, and thou seest the guilty fearing for what is in it, and they say: O woe to us! What a book is this! It leaves out neither a small thing nor a great one, but numbers them (all), and they find what they did confronting them. And thy Lord wrongs not any one. [18:49]*

{وَوُضِعَ الْكِتَابُ فَتَرَى الْمُجْرِمِينَ مُشْفِقِينَ مِمَّا فِيهِ وَيَقُولُونَ يَا وَيْلَتَنَا مَالِ هَذَا الْكِتَابِ لَا يُغَادِرُ صَغِيرَةً وَلَا كَبِيرَةً إِلَّا أَحْصَاهَا وَوَجَدُوا مَا عَمِلُوا حَاضِرًا وَلَا يَظْلِمُ رَبُّكَ أَحَدًا} [الكهف:49]

Also, we read:

> *That Day shall you be brought to Judgment, not a secret of you will be hidden. [69:18-20]*

{يَوْمَئِذٍ تُعْرَضُونَ لَا تَخْفَى مِنكُمْ خَافِيَةٌ (18) فَأَمَّا مَنْ أُوتِيَ كِتَابَهُ بِيَمِينِهِ فَيَقُولُ هَاؤُمُ اقْرَءُوا كِتَابِيَهْ (19) إِنِّي ظَنَنتُ أَنِّي مُلَاقٍ حِسَابِيَهْ} [الحاقة:18-20]

It is remarkable to see that the Qur'an points to one's deeds being recorded in a book. This clearly points to the preservation of information as the universe collapses into a black hole and also points to the presentation of this information in a panoramic scene where no secret is hidden. What an astounding correlation between the message of the Qur'an and the findings of modern science!

Tunneling the Universe

If the current theory of wormholes is true, and if all the black hole singularities are joined together (as is well-expected for the fate of the universe), then there is a possibility that the content of the universe may also tunnel into another universe where a different type of construction is made possible. This only depends on the existence of a certain type of consciousness within living creatures. Theoretically, a wormhole forms when two singularities of two black holes meet to form a throat. This is depicted in Figure 19 below.

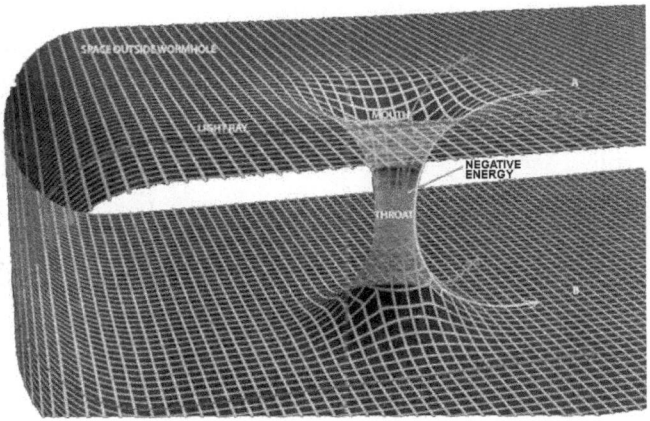

Fig. 19 Wormhole connecting two distant worlds

However, this throat remains to be a bridge (sometimes called the Einstein-Rosen bridge) connecting two regions of the same world. Nevertheless, there can be other formations connecting worlds of different kinds. Theoretical physics is unable as of yet to give us a detailed picture of such constructions; however, it remains a possibility. This is what allows one to view the existence of the other world, beyond what is possible to imagine by our formal physical existence.

Therefore, modern cosmology is open to many possibilities but the most striking thing that we learn is that the description given in the Qur'an regarding the hereafter seems to be feasible within these possibilities. The main difficulty in accommodating the Qur'anic viewpoint is the difficulty in realising any kind of physical existence for the theological and

The Divine Word and the Grand Design

metaphysical concept of the soul. This delays the acceptance of the metaphysical descriptions given in the Qur'an. However, I find that such a dilemma can be partially solved once we realise that the soul as described in the Qur'an is but a collection of identified information. It is composed of the memories that are preserved in unknown records. The difficulty lies in finding where and how such information is recorded and by what means. The Qur'an tells us that Allah is the One who will present this information on doomsday. The problem is that if the recording of such memorial information is done within the body of human being, then it is well-expected that such information will be dispersed throughout space by the time of doomsday. However, if the Creator re-creates the bodies, collecting every atom and every related photon of energy into a new construction, then it would be possible to retrieve the information and have it exposed in the holographic presentation.

It could be that the records are preserved in the metaphysical world of Allah by metaphysical means, as connoted in the Qur'an:

> *Or do they think that We hear not their secrets and their private counsel? (Yes We do) and Our Messengers (appointed angels in charge of mankind) are by them, to record. [43:80]*
>
> {أَمْ يَحْسَبُونَ أَنَّا لَا نَسْمَعُ سِرَّهُمْ وَنَجْوَاهُم بَلَىٰ وَرُسُلُنَا لَدَيْهِمْ يَكْتُبُونَ}
> [الزخرف:80]

The only hint in this verse which allows us to consider a physical means by which information regarding human acts is recorded is the fact that the Messengers of Allah (which are normally understood as the appointed angels) are recording such information. Thus, to my understanding, angels could be understood as a physical means. We need to search for whatever means possible in this respect, including any kind of non-physical channel (like the quantum entanglement channels). This does, however, remain an unsolved problem. Ultimately, we as Muslims submit to this belief, no matter which way our acts are preserved and regardless of whether we find a rational explanation for it or not.

Does the Creation have a Purpose?

Two important issues are involved in discussing this question: the first is the existence of regularities in the natural phenomena which we call 'laws of nature' and the second is to foresee a destiny for the creation. The first issue is a fact that we are accustomed to see, a reality that we rationally conclude through prolonged observations, but the second issue is more of an extrapolation that we may conclude through rational induction.

Laws of Nature and Laws of Physics

Laws of nature have often been confused with the laws of physics by people and even by scientists. While the laws of nature are the actual phenomena occurring, laws of physics are our rational explanations of those laws. History of science and the development of scientific knowledge tell us that our explanations of the laws of nature can drastically change. Einstein's theory of general relativity eclipsed Newton's theory of gravity, although both theories explain the same phenomena, i.e. gravity. While the Newtonian concept of gravity is based on the notion of a field of force emanating from a massive body acting-at-a-distance, Einstein's theory suggested a completely new notion of curved spacetime as a description of gravity. Newtonian gravity is described by one potential producing a three-dimensional force, whereas Einstein's gravity has ten potentials producing curvature of spacetime.

Gravity as a natural phenomenon remains the same. However, since the calculations which are based on Einstein's explanation are more accurate than those obtained based on Newton's theory, Einstein's description is deemed the correct one. This example exposes the difference between laws of physics and laws of nature from which we understand that the latter are immutable laws, whereas the former are not and therefore subject to change.

The Divine Word and the Grand Design

A typical example of this confusion and its consequences is the discussion presented by Paul Davies in his book *The Mind of God* where he uses 'laws of nature' to mean what we would describe as natural phenomena, and 'laws of physics' to point to the phenomena themselves. This kind of confusion may lead to absurdities and to the faulty identification of the entities at play when discussing such vital questions as the creation of the universe in a philosophical context. For example, he claims, "given the laws of physics, the universe can create itself". This is a typical example of what I call confusion or misunderstanding by mixing the two concepts into one common meaning. The belief that the laws of physics are descriptions of natural phenomena was only the case until the beginning of the 20th century, when relativity theory came to replace Newtonian mechanics and his law of universal gravitation with more accurate formulations and a new understanding, and when quantum mechanics uncovered the fact that the laws of classical physics were only an approximate formulation of natural phenomena.

This confusion might have been brought about by the common origin of the words 'nature" and 'physics', as both terms historically expressed the same meaning. The confusion causes misunderstanding over the reality of the laws of physics and leads us to give such laws the status of being in existence 'out there' with exaggerated supremacy and sovereignty. For example, on arguing for the initial conditions of the universe or the laws operating at the initiation of the creation of the universe, Paul Davies suggests that, "Laws of initial conditions strongly support the Platonic idea that the laws are 'out there' transcending the physical universe. It is sometimes argued that the laws of physics came into being with the universe. If that was so, then those laws cannot explain the origin of the universe, because the laws would not exist until the universe existed."[75]

Contrary to the idea presented in this quote, the laws which are actually 'out there' are the laws of nature, with their mathematical construct

---

75 Davies, *The Mind of God*, p.91–92.

or the logic behind their operation not being known to us with absolute certainty. These laws of nature came into being with the universe and we do not know how they could have existed before the birth of the universe.

Richard Dawkins is another example of an author who puts forward speculations drawn from Darwin's theory of evolution and tries to present them as being laws of nature. Whereas evolution *is* a law of nature, being an observed fact, Darwin's theory of biological evolution is not. It could be considered to be a law of biology, however. It is not a problem of mere terminology that I am dealing with here; it runs deeper than that. Thus, I feel the need to clarify the two concepts to enable us to use them accurately in their appropriate contexts.

Indeterminism of Nature

Quantum mechanics teaches us that the laws of nature are indeterministic while we know that the laws of physics are deterministic, being a mathematical construct of what we think is describing the natural phenomena. Being indeterministic, the laws of nature cannot act by themselves. On the other hand, as our mental construct, the laws of physics cannot act by themselves too. So, who puts the 'fire in the equations'?

While it might be acceptable to say that the laws of nature are in the mind of God, it is not acceptable to say that the laws of physics are part of the mind of God. In this context appears the value of the laws of nature as pointing to God, the Creator and Sustainer of the world, in determining the action of these laws. Consequently the world cannot be thought to work without a purpose. However, if we confine our view to see the parts of the material world in isolation, reducing relationships into mere mechanisms, we may not recognise a purpose. This way of viewing the world is in line with the thinking of physicists like Steven Weinberg. But if the world is seen as a whole in every respect and every aspect, we would recognise that there must be a purpose, regardless of

The Divine Word and the Grand Design

how easy or difficult this process is. Herein lies the foresight which leads one to acknowledge the need for God or not, and herein comes the use of the hypothesis of God which neither Laplace nor Weinberg could discover.

> *ALLAH disdains not to give an illustration—as small as a gnat or even smaller. Those who believe know that it is the truth from their Lord, while those who disbelieve say, "What does ALLAH mean by such an illustration?" Many does HE adjudge by it to be in error and many by it does HE guide, and none does HE adjudge thereby to be in error except the disobedient. [2:26]*

{إِنَّ اللَّهَ لَا يَسْتَحْيِي أَنْ يَضْرِبَ مَثَلًا مَا بَعُوضَةً فَمَا فَوْقَهَا فَأَمَّا الَّذِينَ آمَنُوا فَيَعْلَمُونَ أَنَّهُ الْحَقُّ مِنْ رَبِّهِمْ وَأَمَّا الَّذِينَ كَفَرُوا فَيَقُولُونَ مَاذَا أَرَادَ اللَّهُ بِهَذَا مَثَلًا يُضِلُّ بِهِ كَثِيرًا وَيَهْدِي بِهِ كَثِيرًا وَمَا يُضِلُّ بِهِ إِلَّا الْفَاسِقِينَ} [البقرة:26]

There is a built-in curiosity in our rational logic that makes us question the purpose of creation. This curiosity, although it is at different levels of comprehension, is the driving force behind our search for the Creator. Some may be led to deny any role for the Creator, as was the case with the late Stephen Hawking. The reason for such denial is due to the other side of the coin, where we are compelled by our rational thinking to request a cause for every event besides attributing every action solely to that cause.

In fact, logical conclusions alone do not provide us with sufficient comprehension since our construction as conscious beings involves more than pure rationality. Our advanced consciousness enables us to deduce some conclusions that may surpass our rationality. The problem with denying the need for a creator is motivated partly by reductionism, which is the belief that everything and every event can be reduced to material components that provide sufficient reasons for the existence of things and explaining the occurring events. This is by no means true; not unless science reaches its ultimate knowledge about the whole world.

The second reason for not believing in God is the idea that should God exist then the universe would be ruled by miracles. This belief is flawed. On the contrary, if the universe was dominated by miraculous events, then this would indicate the absence of law and order. In such a case we would be unable to identify any sign that leads us to the Creator and Sustainer of the world.

Meeting Stephen Hawking

In August 1977, I met Stephen Hawking during a coffee break at the Eighth International Conference on General Relativity and Gravitation (GR8) held at Waterloo University, Ontario. I asked him: "Do you think, Professor Hawking, that behind all these equations and mathematical formulations we are presenting on the boards of this conference each day, there is something that cannot be described with mathematical equations?" Hawking paused for a while, turning his head slowly from left to right, and said: "If there is something, I believe it has to be logical." Then I asked: "But does your intuition tell you anything about this?" He replied: "I can only say that I am searching for the answer."

After getting the result showing that the universe could have existed for an endless imaginary (not real) time before its physical existence, Hawking exclaimed: "What place, then, for a creator?" In this, he overlooked the fact that imaginary quantities are not directly measurable, despite their role in the mathematical formulations of physics. Investigating the quantum state of the vacuum, Hawking found that the universe could have been created from nothing by gravity alone; accordingly, he claimed in his book *The Grand Design* that there is no need for a creator.

Laurence Krauss

Similar claims were made by the physicist Lawrence Krauss in his book, *A Universe from Nothing*. Both Hawking and Krauss ignore the fact that if the existence of the quantum vacuum is to be considered trivial (since 'nothing' may not need a creator anyway, although necessary to produce real particles out of the quantum vacuum), then the existence of gravity is by no means trivial. Very strong gravity (or spacetime warp) needed to convert nothing into something must have existed in the background and the fluctuations of flat empty spacetime would not have been sufficient to convert the nothing into something.

Confronted with facts that point toward a transcendental existence in a debate with John Polkinghorne, Steven Weinberg declared: "My argument can be falsified if a fiery sword will come from nowhere and hit me for my impiety."[76] In a public lecture, Lawrence Krauss agreed that he might believe in God if one evening he found the stars arranged in the sky to read "I am here". Clearly, both Weinberg and Krauss are implicitly assuming that the existence of God implies that the world is run miraculously, which is not the case.

When Richard Dawkins tried to expand on the hypothesis of a multiverse to refute the pre-setting of a finely-tuned universe and put the question to Steven Weinberg during an interview, Weinberg remarked that one should not underestimate the fix that atheists are in, which is that consistent mathematical results cannot be guaranteed to describe a realistic state, since there are many consistent mathematical formulations that do not find real presence in nature.

This implies that both Professor Weinberg and Professor Krauss could see the necessity for God, but only if the universe were to run miraculously. A miraculously run universe is defined by the absence of any order or law that can explain it. Such a universe may not need God

---

76 Steven Weinberg in a debate with John Polkinghorne at the SSQ symposium "Science and the Three Monotheisms: A New Partnership?", Granada, Spain, 23–5 August 2002.

altogether, but a mere force to sustain the chaos. This is what one would usually expect from blind nature.

The problem with many scientists from the physics community who do not see the need for God is that they would like to see God emerging from one of their equations explaining, say, the masses of elementary particles. Only then would they believe in God. This is perhaps why the Higgs boson has been dubbed the 'God particle'.

Chaos and Order

The counterargument to a chaotically run universe would be to say that an ordered universe, ruled by well-defined laws that do not necessarily lead to fully deterministic results, is what necessitates the existence of God. This is because we need to have an operator for the laws of nature and we need an algorithm by which the actions of these laws are coordinated in order to achieve fruitful results. Otherwise, the action of the laws of nature cannot be explained. An operator of these laws cannot be something that belongs to the same set, for such an operator would have to abide by the same laws and thus could not be a ruler over them. This understanding of the divine agent rules out the notion of the 'God of the gaps', which is a famous accusation against an apologetic argument for the role of divine action in nature.

Let us look at some vital arguments inspired by modern physics that could pose some basic problems for comprehending nature on the theoretical level. Besides the classical problem of the creation of the world, there are a number of problems in front of us that are needed in the investigation. These are:

1. The Problem of the Operation: Classically, it was taken for granted that physical laws, being a set of expressions that describe the relationships between variables of certain phenomena, could operate on their own without the need for an operator. A close look at these laws through quantum theory shows that these

laws are expressed as mathematical operators acting on the state of the system that produces the physical observable (called the 'eigenvalue'). As such, the physical observable is produced by the action of an operator on the state of the system. Mathematically, the operator is a symbol representing some kind of operation, say, an infinitesimal translation. The question concerns how to realise such an operator in the activity of the real world, and who the real world is being operated by.

2. The Problem of Coordination: Physical laws may effectively be contradictory. This is why pure chance is blind. Without coordinating the actions of these laws (operations), the world would be in a mess. Such coordination is made through the inherent condition of validity of the physical law. Here some determinism may seem to be at work, but this is not the case, since the operation of the physical law within the given condition itself is undetermined.

3. The Problem of Belongingness: A global operator/coordinator cannot be part of the world or belong to the same set of physical laws acting within the world; otherwise, such an operator/coordinator would have to abide by the same status of any other law belonging to the set.

4. The Problem of Interaction: If the global operator/coordinator is to be external to the set of physical laws that are at work in the world, the question arises over its form and its means of interaction with the world.

A set of several coordinated operations constitute what I would call an algorithm. This is a set of steps that are needed to carry out a given process and establish the causal joint between the individual steps of that process. The algorithm is predestined, but the outcome may not be so. The interplay between contingency and necessity is what would lead to fruitfulness. Contingency is the primary setup for every item in nature; it is the possibilities, the innate properties acquired by the object, or, as we call it in reductionist terminology, the 'number of degrees of freedom', and necessity is what would make the choice. This is why necessity implies destiny.

A choice made would mean that a *will* would have to be at work. The world does not have a will of its own but does have freedom. This freedom is expressed through the contingency of events and algorithms. It is for this reason that we are in need of a global operator/coordinator that will take care of all operations, which must, as I have pointed out, be inherently predestined.

The world contains much that we are yet to discover. Our thirst and pursuit for discovery continues, and for this reason, it would be incorrect to deny basic facts of our logical construct, i.e. that everything which begins to exist must have a cause, hence, the universe began to exist so it must have a cause. Science can never provide us with the ultimate cause and as a result, we have to accept that the ultimate cause is transcendent. Many signs and simple evidence from the world we live in tell us that such a world cannot exist without a creator, nor can it be sustained without a sustainer.

Life in Outer Space

During the past quarter of a century, astronomers have discovered a large number of planets which belong to the many extra-solar systems in our Milky Way. It is estimated that there are more than 200 billion stars in our galaxy, so we would expect them to harbour hundreds of billions of exoplanets. Accordingly, a controversial question has been raised: are any of these planets able to accommodate life? In order to answer this question, many researchers have studied and defined the necessary conditions that make a planet habitable. Data obtained from interstellar dust and space observatories observing the recently discovered exoplanets has been analysed for any indication that they would be suitable for nurturing life. There are a number of crucial factors which constrain the possibility of life on other planets, the most important of which are:

1. Having a solid, stable surface.
2. Having sufficient quantities of liquid water on the planet.
3. Having an atmosphere which contains suitable volumes of gases required for establishing and sustaining life there. The volume as well as the structure of the atmosphere should be adequate to protect the surface of the planet from any dangers coming from outer space.
4. To have magnetic shielding with enough strength to protect the planet against the dangers of stellar winds produced by the stars.

Aside from these points noted above, there are in fact many other requirements for sustaining life on extra-solar planets and, as we see in our solar system, certain planets could have accommodated life at some time in their history (like Venus and Mars) but the lack of one or more essential factors means that the chances of supporting life are no longer feasible. The conditions required to establish and sustain life in any place in the universe is very difficult due to one major aspect—there are many dangers that can eliminate life, whilst the establishment of life, including the more primitive ones, necessitates a lengthy period of time. Several planets which are similar to the Earth in size and mass have been discovered but there is still debate as to whether or not they are qualified for accommodating a flourishing life form.

Philosophically, the possibility of life on other planets is an important topic for study in the new kalam. In daqiq al-kalam, the physical and biological conditions along with their implications are assessed, while in jalil al-kalam, it is the social, moral and religious questions of such a situation that are discussed. A most interesting situation would arise if we were to encounter a creature much more advanced than us, creatures that developed, say, millennia before us. Of course, it may be that there is some sort of a natural constraint on the age of advanced civilisations, such that no peoples can progress to a super stage of development before being destroyed. This might well be the case if the example we are experiencing on planet Earth is typical of what is to be found in other places in the universe.

Chapter Nine

Problematic Verses

As mentioned earlier, the Qur'an states that some of its verses are dubious or unclear (mutashābeh), as shown in verse [3:7]. My explanation for the existence of these verses is that since the Qur'an is the word of Allah and emanates from His absolute knowledge, its wording has to be formulated in such a way as to express facts precisely. People who receive the Qur'an throughout the ages do not always possess the most accurate knowledge or know actual facts; knowledge is continuously developing. For example, at the time the Qur'an was revealed, the commonly-held belief was that the earth is the immovable centre of the universe and that the Sun, the Moon and all the other planets and stars are rotating around it. How would it be feasible then to express the fact that the Earth and the other planets rotate around the Sun, and state that the Earth makes a full rotation around its axis in one day? If the Qur'an stated such a fact explicitly, nobody would believe it.[77] People would claim that the Qur'an is nonsensical. This is one example of many that would have certainly contributed negatively to the image of Islam and would have

---

[77] There is a narration accorded to the Prophet which states, "We prophets have been ordered to talk to people at the level of their comprehension." This hadith is narrated from different sources with different levels of authenticity.

made the propagation of its message very difficult, if not impossible. For this reason, details regarding the creation and nature were conveyed using expressions which formally agreed with the prevailing knowledge of the day, whilst simultaneously corresponding with facts discovered later. This explanation is supported by the careful and often peculiar choice of wording in the Qur'an.

This situation therefore requires us to delve into the meanings of the words as given in Arabic lexicons in addition to considering the context in which the word is used. Moreover, we have to look closely for the same words appearing in other contexts in the Qur'an. Obviously, scrutiny of the text would not be sufficient unless one has specialized knowledge from affirmed related scientific facts and conclusions. Hence, the question of the dubious verses in the Qur'an has no clear-cut answer and for this reason, the Qur'an itself stresses in the verse [3:7] that those who have profound knowledge of science would be able to appreciate its content.

In the following pages, I will discuss a few problematic issues from verses in the Qur'an and present my own views. Of course, this discussion is by no means exhaustive. There are many other problematic issues on which I could not present firm opinions and which had to be left for the time being until more scientific discoveries can shed some light on the content. Remember, the point is not to forge scientific discoveries and claim that they were expressed in the Qur'an 1450 years ago, but to recognise that the development of our knowledge about the world is in conformity with what the Qur'an tells us.

The Heavens

The Qur'an does not provide a continuous textual description of the creation, as is the case with the Old Testament. Instead, there are passages scattered throughout the Qur'an which deal with certain aspects of the creation. These provide partial information on successive events

Problematic Verses

and mark their development with varying degrees of detail. To obtain a clear idea of how these events are presented, the textual fragments—scattered throughout a large number of chapters of the Qur'an—have to be brought together.

In fact, this dispersal is a general characteristic of the Qur'an where many important subjects are treated in the same manner; earthly or celestial phenomena, or problems concerning man that are of interest to scientists. Even stories of the prophets who preceded the Prophet Muhammad ﷺ are told in the same way. This characteristic makes it very hard to draw conclusions from a hermeneutical analysis of the Qur'an.

The word 'heaven' (single) is repeated in the Qur'an 120 times and the word 'heavens' (plural) is repeated 190 times. In total, the two words are repeated 310 times. A few years ago my colleague, Dr. M. Al-Zu'bi, Professor of Arabic, and I conducted a study investigating the meaning of these words as they appear in different verses of the Qur'an. The results showed that they have several meanings; some are very clear, and others are not so clear. In all verses the meaning of these words were found to be context-dependent. However, we were still able to differentiate these meanings into several groups.

In one context, heaven means the sky, which is the extended space above our heads as it is in [6:125] and [15:14], for example. In another context, the word heaven means the clouds as in [2:22]. In the third context, heaven means the atmosphere and the space near the Earth, as in verse [21:32] in which it is said that the heaven is ordained to be a "guarded roof". Also, the verse [81:11] points to the removal of the atmosphere as the Sun becomes a red-giant and approaches the Earth. In the fourth context, the term heavens points to the solar system, as in [41:12]. From this the seven heavens are interpreted as the seven celestial spheres or orbs of the planets. In the fifth context, the word heaven may point to the whole universe as it is in [51:36] where we are told that heaven is continuously expanding, and in [21:104] which states that heaven will collapse on doomsday.

The Divine Word and the Grand Design

The main conclusion that we can draw from this is that the terms 'heaven' and 'heavens' have meanings which are distinct from that in the Old Testament. For example, the Old Testament tells us that God created heaven from the water, where He separated the water by the firmament into two parts; the upper part was called heaven and the lower part was kept on the Earth. Conversely, the Qur'an tells us that the Earth and heaven were joined together forming one unit of creation and they were parted.

> *Have not those who disbelieve see that the heavens and the Earth were joined together as one united piece, then We parted them?*
> *[21:30]*

{أَوَلَمْ يَرَ الَّذِينَ كَفَرُوا أَنَّ السَّمَاوَاتِ وَالْأَرْضَ كَانَتَا رَتْقًا فَفَتَقْنَاهُمَا وَجَعَلْنَا مِنَ الْمَاءِ كُلَّ شَيْءٍ حَيٍّ أَفَلَا يُؤْمِنُونَ} [الأنبياء:30]

The Qur'an also tells us that the heavens were formed out of smoke:

> *Then He turned towards the heaven when it was smoke, and said to it and to the Earth: "Come both of you willingly or unwillingly." They both said: "We come, willingly." So He ordained them seven heavens in two days, and revealed in every heaven its affair.*
> *[41:11]*

{ثُمَّ اسْتَوَى إِلَى السَّمَاءِ وَهِيَ دُخَانٌ فَقَالَ لَهَا وَلِلْأَرْضِ ائْتِيَا طَوْعًا أَوْ كَرْهًا قَالَتَا أَتَيْنَا طَائِعِينَ} [فصلت:11]

The verse [21:30] has been understood by some contemporary authors who have written on the 'scientific miracles of the Qur'an' as pointing to the creation of the universe in a Big Bang. However, once we study the verse carefully in conjunction with other verses, we can recognise that the verse may have nothing to do with the Big Bang. The fact that the Sun and the Earth and the other planets of the solar system contain heavy elements including uranium suggests that the solar system, including the Earth, was formed out of the remnants of large nebula left over from an ex-generation large star, believed to have ended its life in a huge supernova explosion billions of years ago. This becomes clear when we analyse the verses regarding the seven heavens in the following section.

The Seven Heavens

The Qur'an mentions that Allah formed seven heavens that are stacked one above the other:

> Who has created the seven heavens one above another, you can see no fault in the creations of the Most Beneficent. Then look again: Can you see any rifts? [67:3]

{الَّذِي خَلَقَ سَبْعَ سَمَاوَاتٍ طِبَاقًا مَا تَرَى فِي خَلْقِ الرَّحْمَنِ مِنْ تَفَاوُتٍ فَارْجِعِ البَصَرَ هَلْ تَرَى مِنْ فُطُورٍ} [الملك:3]

As for the meaning of the seven heavens, it seems that this concept is as old as the major faiths that appeared since the dawn of history. The concept appears in the ancient Mesopotamian religions, as well as in Judaism and Hinduism, indicating that it is ingrained in traditional religious ideas of creation. In my opinion, the concept of the seven heavens remains unclear at the present time, unless we consider it to represent the seven celestial spheres.

The Seven Earths

The idea of the seven Earths has been construed from the following verse:

> It is Allah Who has created seven heavens and of the Earth the like thereof (i.e. seven). His Command descends between them, that you may know that Allah has power over all things, and that Allah surrounds (comprehends) all things in (His) Knowledge. [65:12]

{اللَّهُ الَّذِي خَلَقَ سَبْعَ سَمَاوَاتٍ وَمِنَ الْأَرْضِ مِثْلَهُنَّ يَتَنَزَّلُ الْأَمْرُ بَيْنَهُنَّ لِتَعْلَمُوا أَنَّ اللَّهَ عَلَى كُلِّ شَيْءٍ قَدِيرٌ وَأَنَّ اللَّهَ قَدْ أَحَاطَ بِكُلِّ شَيْءٍ عِلْمًا} [الطلاق:12]

Since the very early times of Qur'anic exegesis, commentators have held widely differing opinions about the meaning of the seven earths. Some have considered it to mean seven planets situated far away in space on which there are creatures similar to man in every respect, and may even have better morals. Others maintain that they point to the seven layers forming the body of the Earth, but this has been objected to on the basis that there are only three known layers. The third interpretation in the tradition is that the seven earths point to the seven geographical regions of the Earth.

The seven heavens may even be understood to point to the seven celestial spheres known at the time of the Prophet Muhammad ﷺ from ancient astronomy, supported by the fact that the Qur'an also mentions seven earths. Now it is known that there are five solid planets of solid crust: Mercury, Venus, Earth, Mars, Pluto and two other dwarf planets. Here we should be careful and conclude that the concept of the seven earths remains dubious.

The idea that the heavens are formed out of smoke is in line with the findings of astronomical observations and the geological formation of the planets. However, the Qur'an seems to indicate in some verses that the Earth was formed before the seven heavens. This is somewhat doubtful, though not impossible. No one knows which planet was formed first. It could be that the Earth nucleated first and the other planets were formed subsequently. Incidentally, the verses in the Qur'an which allude to the Earth's earlier formation suggest that the formation of the seven heavens coincided with the formation of the Earth. This is indeed scientifically sound.

> *Say: Do you verily disbelieve in Him Who created the Earth in two Days and you set up rivals (in worship) with Him? He placed therein firm mountains from above it, and He blessed it, and allocated therein its sustenance in four Days* [41:9–10].

Problematic Verses

{قُلْ أَئِنَّكُمْ لَتَكْفُرُونَ بِالَّذِي خَلَقَ الْأَرْضَ فِي يَوْمَيْنِ وَتَجْعَلُونَ لَهُ أَندَادًا ذَلِكَ رَبُّ الْعَالَمِينَ (9) وَجَعَلَ فِيهَا رَوَاسِيَ مِنْ فَوْقِهَا وَبَارَكَ فِيهَا وَقَدَّرَ فِيهَا أَقْوَاتَهَا فِي أَرْبَعَةِ أَيَّامٍ سَوَاءً لِلسَّائِلِينَ} [فصلت:10-9]

It might be a fruitless endeavour to investigate the meaning of the seven heavens beyond what we can deduce from the space beyond the Earth's atmosphere.

Is Heaven Built?

There are a significant number of verses in the Qur'an which state that heaven has been built or is a building. This can be found in the following verses: [91:5], [51:47], [40:64] and [79:27]. The understanding that heaven is a built wall may be supported by other verses which tell us that heaven has gates [7:40], [15:14], [54:34] and [78:19]. In another verse, if translated literally, the Qur'an tells us heaven is a guarded roof [21:34]. Furthermore, in some verses we are presented with the image of heaven becoming cracked and torn up on doomsday like in [72:18], [82:1] and [84:1]. It was therefore not unusual to find Muslim artists depicting this image in their works.

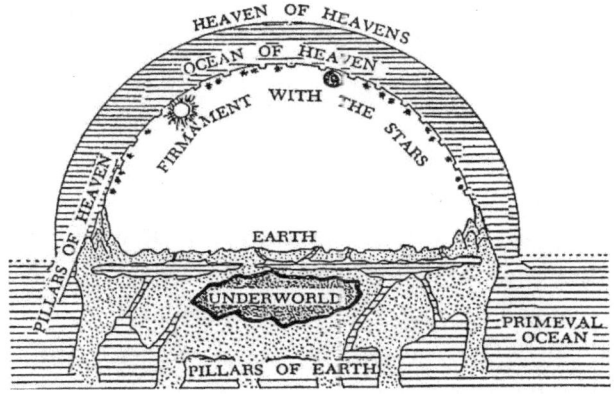

Fig. 20 An image of the heavens

The Divine Word and the Grand Design

In fact, the description of heaven in the Qur'an falls within the context of what I call the dubious (mutashābeh) verses. The notion of heaven as a roof and as a built wall was part of the common perception of the public at the time of revelation. For this reason, the Qur'an uses this image in a number of verses. However, the Qur'an does not necessarily prompt us to take this description literally, as becomes evident when we analyse other verses in which the word heaven is mentioned. The strategy here is to recognise the discrepancy between the content of these verses, which implies that heaven cannot be considered to be built like a wall but as a structure. This can be seen in the verse:

> Blessed is He who has established constellations in the sky and made therein a lamp and a shining Moon. [25:61]

{تَبَارَكَ الَّذِي جَعَلَ فِي السَّمَاءِ بُرُوجًا وَجَعَلَ فِيهَا سِرَاجًا وَقَمَرًا مُنِيرًا}
[الفرقان:61]

Clearly the presence of the constellation, stars, Sun and the Moon in the heaven indicates that it is a complicated structure which has to be built. For this reason the Qur'an states that the creation of heavens is even more complicated that the creation of man.

> The creation of the heavens and the earth is indeed greater than the creation of mankind, yet most of mankind know not. [40:57]

{لَخَلْقُ السَّمَاوَاتِ وَالْأَرْضِ أَكْبَرُ مِنْ خَلْقِ النَّاسِ وَلَكِنَّ أَكْثَرَ النَّاسِ لَا يَعْلَمُونَ}
[غافر:57]

In fact this verse has a deeper meaning in that it is implicitly pointing to the fact that the Heavens has been created out of nothing while man has been created out of clay which was formed after the creation of the heavens and the Earth. Note that in the tailing of this verse it is mentioned *'yet most of mankind know not,'* asserting that this piece of information is not known to most people.

Flatness of the Earth

Influenced by an international movement which claims that the Earth is flat, a minority of Muslims found that this claim is supported by verses from the Qur'an. Non-believers on the other hand, seized this opportunity to claim that the Qur'an contradicts science by telling us that the Earth is flat.

As I mentioned earlier, the word Earth carries different meanings in the Qur'an. In one context it is a piece of limited land, in another it is the state or kingdom, and in another it is the Earth as a planet. Nowhere in the Qur'an does it explicitly state that the Earth is flat, despite several verses which can be taken at face value to imply that it is. But which meaning is given to the word Earth in those verses? Is it the piece of land or is it the whole Earth?

> *And the Earth! We have spread it out, and set thereon mountains standing firm, and have produced therein every kind of lovely growth (plants). [50:7]*

{وَالْأَرْضَ مَدَدْنَاهَا وَأَلْقَيْنَا فِيهَا رَوَاسِيَ وَأَنْبَتْنَا فِيهَا مِنْ كُلِّ زَوْجٍ بَهِيجٍ} [ق:7]

The essential word *madadnāhā* means extended it or spread it, which does not necessarily denote that the Earth is flat.

> *And by the Earth and Him Who spread it. [91:6]*

{وَالْأَرْضِ وَمَا طَحَاهَا} [الشمس:6]

All English translations express the sensitive word *taḥāhā* to mean spreading or extending it, but one finds that in traditional exegesis of the Qur'an the apparent meaning was adopted, aligning with the common belief that the Earth is flat. Classical Arabic lexicons point to double meanings for taḥā: one which implies extending and flattening, the other meaning round. The meaning which implies roundness is supported by the famous lexicon of Ibn Manzur entitled *Lisan al-Arab*, a highly respected reference book in which he says that the word taḥā means to 'extend and flatten.' He goes on to say that Ibn Sīdah, another highly reputed lexicographer, stated that tahā is also used in connection with describing a huge umbrella. As such, we can understand that the use of the word tahāhā in the above verse is seemingly deliberate. The use might be intended to describe the local flatness of the Earth which is a consequence of the extension of its size; otherwise, it is round.

> *It is God who extended out the Earth and fixed mountains and placed rivers therein. He made a pair of every fruit and made the night cover the day. All this is evidence (of the existence of God) for the people who think. [13:3]*

{وَهُوَ الَّذِي مَدَّ الْأَرْضَ وَجَعَلَ فِيهَا رَوَاسِيَ وَأَنْهَارًا وَمِنْ كُلِّ الثَّمَرَاتِ جَعَلَ فِيهَا زَوْجَيْنِ اثْنَيْنِ يُغْشِي اللَّيْلَ النَّهَارَ إِنَّ فِي ذَلِكَ لَآيَاتٍ لِقَوْمٍ يَتَفَكَّرُونَ}[الرعد:3]

The word *madda* in Arabic means extended, out of which the word *imtidād* (meaning 'extension') is derived. The verse also mentions the succession of the day after the night and asks people to contemplate on it.

In the same vein, there is another verse which might be considered the strongest explicit indicator of the flatness of the Earth:

> *And Allah has made the Earth for you as a carpet (spread out). That ye may go about therein, in spacious roads. [71:19–20]*

{وَاللَّهُ جَعَلَ لَكُمُ الْأَرْضَ بِسَاطًا (19) لِتَسْلُكُوا مِنْهَا سُبُلًا فِجَاجًا} [نوح:19–20]

In this verse the Qur'an tells us that the Earth is made flat like a carpet for us so that we may go about therein in spacious roads. This clearly indicates that the flattened parts are localised regions and not the whole Earth, since the verse speaks of our travelling through open places. Hence, the earth in this context indicates the land.

Motion of the Sun

Does the Earth rotate around the Sun or does the Sun rotate around the Earth? Some people claim that the Qur'an supports the idea that the Sun is rotating around the Earth, not the other way round. This might be true since to the best of my knowledge the Qur'an never mentions the rotation of the Earth around the Sun, either explicitly or implicitly. We must remember that here we are discussing motion, which is relative depending on the frame of reference of the observer. For an observer on the surface of the Earth, the Sun makes a full trip around the Earth in 24 hours. For a hypothetical observer located near the Sun, the Earth rotates around the Sun in 365 days. So when we read in the Qur'an that the Sun and the Moon are swimming in their orbs, such a statement is in no contradiction with scientific facts. The Sun has its relative motion with respect to the centre of the galaxy and the Moon is moving around the Earth. Incidentally, if you are standing on the surface of the Moon you will think that the Earth is moving around the Moon and not the other way round. Again, this is because motion is relative.

The Throne and the Water

A classic problem in Qur'anic exegesis concerns a verse which speaks of the Throne of Allah being on water:

> *And He it is Who has created the heavens and the earth in six Days and His Throne was on the water, that He might try you, which of you is the best in deeds. [11:7]*

{وَهُوَ الَّذِي خَلَقَ السَّمَاوَاتِ وَالْأَرْضَ فِي سِتَّةِ أَيَّامٍ وَكَانَ عَرْشُهُ عَلَى الْمَاءِ لِيَبْلُوَكُمْ أَيُّكُمْ أَحْسَنُ عَمَلًا} [هود:7]

Clearly the context is the creation of the heavens and the Earth. We also notice that the reason behind creation is to test which of us is the best in deeds. In between comes the statement: *"and His Throne was on the water"*. The key word in this verse is the Throne. What can be the

The Divine Word and the Grand Design

Throne of Allah in this verse other than life itself? So, the simple and straightforward understanding of the verse is that Allah has based His creation on water—the most fundamental element for life—in order to try us and see whether we perform good deeds. Otherwise there is no point in bringing the statement in the context of the Earth's creation and specifically the creation of man, who seems to be the whole purpose of the former. This understanding is further supported by the verse:

> *Have not those who disbelieve known that the heavens and the earth were joined together as one united piece, then We parted them? And We have made from water every living thing. Will they not then believe? [20:30]*

{أَوَلَمْ يَرَ الَّذِينَ كَفَرُوا أَنَّ السَّمَاوَاتِ وَالْأَرْضَ كَانَتَا رَتْقًا فَفَتَقْنَاهُمَا وَجَعَلْنَا مِنَ الْمَاءِ كُلَّ شَيْءٍ حَيٍّ أَفَلَا يُؤْمِنُونَ} [الأنبياء:30]

Here, pointing to the dependence of life on water and the related creation of the heaven and the Earth is quite clear.

It seems that the dispute amongst traditional commentators regarding this point was ultimately not worth it. The problem, as far as I can see, arises from overlooking the metaphor in the verse.

Timescale of the Creation

In several verses which fall under the category of dubious verses, the Qur'an states that the heavens were created in six days. This is in line with the narrative expressed in the Old Testament; however, the Qur'an asserts that Allah created the heavens and the Earth with truth, or truthfully. On analysing these verses, we can see that the words 'truth' and 'truthfully' point to the existence of a reference or a measure. Such a measure refers to a law that forms a legal reference for whatever is to be considered just or truthful. If the creation was done miraculously without any verifiability, then the universe would not abide by any sense of order. However, the creation was ordered and had to be made in accordance with a sequence and processes which take time and satisfy certain conditions and requirements. 'Days' of the creation can be understood to signify the chapters of the creation. Indeed, this is what has occurred over the history of the development of the universe.

Thus, the creation timescale given in the Old Testament and the Qur'an simply conveys that the creation occurred in several stages. The main issue in this kind of construction and in the existence of laws is that behind such a ruling, there is an order pointing to the One who is putting fire into the equations and describing the natural laws and processes. He, the Creator, the Sustainer and the Ultimate of all knowledge and existence, is the only meaningful existence deserving to be worshipped.

Epilogue

In this book I attempted to advance the view that the Qur'an, a Holy book presenting sacred statements emanating from the eternal knowledge of the Creator, is an invariably unique text. As such, its rich, multiple layers allow for different levels of understanding corresponding with the development of knowledge over time. The structural and contextual character of the Qur'an is required to understand the perpetually living word of the Divine being expressed in human formal language. Arabic is among the richest languages in the world and through the wealth of semiotics it contains, is capable of bearing the word of Allah across the ages. This is the main reason for Arabic being the chosen language to communicate the message of the Qur'an:

We verily, have made it a Qur'an in Arabic that you may be able to comprehend. [43:3]

{إِنَّا جَعَلْنَاهُ قُرْآنًا عَرَبِيًّا لَعَلَّكُمْ تَعْقِلُونَ} [الزخرف:3]

So, the Qur'an is not a mere revelation for missionary purposes. It is a scripture containing a worldview, encompassing everything from religious instructions to descriptions of natural characteristics, all of which

The Divine Word and the Grand Design

support its authentication from a divine source. This is the main aspect that renders the Qur'an a perpetual miracle.

Interpreting the Qur'an thus requires competence in specialized knowledge of natural sciences as well as acquaintance with Arabic expressions and metaphors; both are essential for deciphering its messages and the knowledge it contains. It is therefore no accident that the Qur'an addresses people of knowledge and wisdom in some verses, and that we find the following statement:

> *We will show them Our Signs in the universe, and in their own selves, until it becomes manifest to them that this (the Qur'an) is the truth. Is it not sufficient in regard to your Lord that He is a Witness over all things? [41:53]*

{سَنُرِيهِمْ آيَاتِنَا فِي الْآفَاقِ وَفِي أَنْفُسِهِمْ حَتَّى يَتَبَيَّنَ لَهُمْ أَنَّهُ الْحَقُّ أَوَلَمْ يَكْفِ بِرَبِّكَ أَنَّهُ عَلَى كُلِّ شَيْءٍ شَهِيدٌ} [فصلت:53]

But does this mean that we can deduce science about ourselves and about nature from the Qur'an? The short answer is no. The Qur'an is not a book of science, but can be taken as a guide directing us to investigate some areas of interest and can be considered a source for ethical codes. Such codes will help us deal with nature and society in a way that maintains the sustainability of individual happiness and social welfare.

One of the main goals of religious teaching in Islam is to harmonise people with nature. The spiritual character of the Qur'an resonates with the people who understand its content. This is attained by a deep understanding of the meanings and becoming in harmony with the rest of the world. Such a status is achieved whenever you feel that you are an integrated part of the world and that you are performing your acts on the highest level of conformity with other creatures. This kind of experience has been seen in reflections on the Qur'an by scholars of Islam, especially the verses which take the reader through a journey of wonders and a feeling of oneness with nature, knowing that we are being sustained by one Creator. This kind of understanding produces a feeling of balance, justice and happiness within us, giving us high mental stability and confidence and releasing us from the pressures of everyday life. Once understood, the Qur'an is a great source of enlightenment for humanity and the perfect guide on how to lead a fulfilling life.

Index

A

Al-Attas 30
Al-Ghazālī 36
Al-Khalili 142, 144
Al-Rāzī 149
Al-Zubi 177
Antarctic 73
Anthropic principle 100, 101, 102, 103, 112, 114, 115, 116, 117, 118
Arabic 1, 3, 6, 11, 14, 27, 28, 29, 33, 49, 61, 64, 80, 85, 126, 127, 129, 134, 177, 183, 184, 189, 190
Aristotle 125, 147
Atmosphere 49, 53, 54, 59, 66, 67, 68, 70, 73, 75, 77, 147, 172, 177, 181
Atomism 28, 35

B

Barrow 102, 103, 105, 114
Big Bang 93, 154, 155
Black hole 83, 159, 160

C

Carbon 45, 48, 49, 63, 66, 67, 81, 96, 97, 101, 103, 117, 141, 156
Causality 17, 26, 36, 143
Cerebrospinal Fluid 18
Cloud Chamber 92
Clouds 69, 71, 73, 177
Consciousness 7, 16, 17, 18, 23, 88, 89, 99, 114, 115, 116, 161, 166
Contingency 138, 171
Cosmic Microwave Background 156
Cosmological Doctrines 24, 25
Creation 3, 4, 12, 15, 16, 26, 34, 35, 45, 61, 65, 69, 70, 89, 90, 91, 97, 98, 99, 102, 104, 108, 111, 112, 113, 114, 115, 116, 117, 118, 122, 125, 126, 127, 129, 130, 131, 133, 135, 140, 146, 147, 157, 158, 163, 164, 166, 169, 176, 178, 179

D

Darwin, Charles 122, 123, 124, 138, 144, 146, 165
Davies 93, 98, 101, 102, 103, 128, 164
Dawkins 104, 105, 145, 165, 168
Descartes, Rene 16
Determinism 26, 139, 143, 170
Discreteness 35
DNA 96, 116, 137, 140, 143, 144
Doomsday 46, 52, 58, 84, 88, 157, 162, 177, 181
Dubious Text 10, 11, 175, 176, 180, 182, 186

E

Einstein 8, 17, 32, 87, 100, 106, 151,

154, 158, 161, 163
Einstein-Rosen bridge 161
Entanglement 17, 162
Eternity 16
Everett 106
Evolution 97, 103, 114, 121, 122, 123, 124, 127, 128, 129, 132, 134, 135, 136, 137, 138, 139, 140, 141, 142, 143, 145, 146, 155, 165
Expansion of the universe 153, 154

F

Face of the Moon 57
Faith 9, 12, 13, 15, 16, 18, 23, 30, 88, 122, 136, 179
Falsifiability 9
Falsifiability of Science 9
Fine tuning 101, 103, 104, 110
Flatness of the Earth 64, 183, 184
Friedman 151

G

Galileo 83, 137
Gamow 154, 155, 156
Genes 140, 144
Global Warming 67
God-of-the-gaps 8
Gravity 44, 47, 54, 60, 79, 81, 82, 83, 98, 153, 159, 163, 167, 168
Greek civilization 26
Guth 108

H

Hawking 159, 166, 167, 168
Heaven 11, 177, 178, 181
Helium 45, 46, 48, 52, 66, 79, 81, 94, 95, 96, 155, 156

Hubble 150, 151
Hydrogen 45, 46, 47, 49, 66, 79, 80, 94, 96, 98, 155, 156

I

Ibn al-Shātir 148
Ibn Faris 49, 134
Ibn Kathīr 90, 129, 132
Ibn Manzūr 183
Ibn Rushd 33
Ibn Sīdah 184
Ikhwan al-Safa 148
Islamic art 28
Islamization of Knowledge 37

J

Jesus 1, 5, 6, 125
Jupiter 44, 60, 148

K

Ka'b Al-Aḥbār 122
Krauss 65, 168

L

Laws of Nature 163, 165
Laws of Physics 163, 165
Linde 106, 108
Lunar month 50, 55, 59

M

Magnetic poles 73
Magnetic Poles 73
Magnetic shield 59, 72, 172
Mars 60, 74, 147, 148, 172
McFadden 140, 141, 142, 144
Messengers 1, 134, 162
Meteorites 54, 63, 68, 70, 71, 74
Mind 2, 6, 8, 16, 115, 116, 122, 134,

151, 164, 165
Miracles 5, 66, 143, 167, 178
Moses 1, 5
Motion of the Sun 46
Muhammad 1, 2, 3, 5, 6, 10, 12, 14, 19, 20, 21, 22, 30, 34, 38, 51, 64, 65, 71, 72, 77, 122, 132, 177, 180
Multiverse 103, 105, 106, 107, 108, 109, 110, 140, 141, 142, 168
Muragha 148
Mutakallimūn 28, 36, 115
Mutations 123, 124, 137, 138, 139, 142, 143, 144, 145

N

Nasr 25, 28, 29
Nature 4, 5, 7, 8, 15, 16, 25, 26, 27, 29, 31, 36, 56, 66, 91, 100, 102, 104, 106, 124, 127, 128, 134, 137, 138, 139, 143, 146, 163, 164, 165, 168, 169, 171, 190
Necessity 104, 126, 138, 139, 168, 170
Neutrino 92, 93, 94
Neutron star 82, 83
Newton 137, 163
Nūr 13, 14

O

Oscillatory universe 157
Oxygen 45, 66, 68, 96, 97, 156

P

Pluto 60, 180
Popper 9
Probability 31, 92, 97, 99, 101, 128, 138, 139

Pulsar 85
Purpose 3, 7, 14, 40, 99, 104, 105, 110, 114, 115, 117, 118, 129, 131, 135, 136, 139, 141, 145, 163, 165

Q

Quantum mechanics 31, 104, 106, 109, 139, 140, 144, 164

R

Re-creation 35, 140
Red-Giant 48
Rees 105
Revelation 13, 24, 25, 27, 69, 71, 72, 154, 182

S

Saturn 44, 60, 147, 148
Schrodinger 139
Shari'a laws 126
Smolin 142
Space 17, 31, 35, 36, 50, 66, 68, 70, 72, 76, 77, 91, 95, 106, 108, 150, 154, 155, 157, 158, 159, 162, 171, 172, 177, 180, 181
Spacetime 32, 106, 107, 108, 154, 159, 163, 168
Staune 137
Supernova 82
Sustainer 1, 15

T

Taqwūr 46, 54
Taskhīr 110
Tawhid 2, 12, 23
Tegmark 107, 108, 109
Temporality 34

Time 89, 108
Turok 108

U

Unicity 26, 27, 28
Universe 7, 8, 11, 12, 16, 18, 32, 35,
　　　　41, 53, 54, 59, 60, 66, 75, 79,
　　　　80, 83, 91, 92, 93, 94, 95, 96,
　　　　97, 98, 99, 100, 101, 102, 103,
　　　　104, 105, 106, 107, 108, 109,
　　　　110, 112, 113, 114, 116, 117,
　　　　118, 122, 126, 127, 128, 136,
　　　　141, 143, 146, 147, 150, 151,
　　　　152, 153, 154, 155, 156, 157,
　　　　158, 159, 160, 161, 164, 165,
　　　　167, 168, 169, 171, 172, 173,
　　　　175, 177, 178, 190
Uranus 44, 60

V

Van Allen Belts 73
Vilenkin 107

W

Weinberg 65, 101, 105, 165, 168
Wheeler 102, 106, 114
White Dwarf 49, 51
Wormholes 160

www.ingramcontent.com/pod-product-compliance
Lightning Source LLC
Chambersburg PA
CBHW031954080426
42735CB00007B/384